FACETS OF FAITH AND SCIENCE

Volume 3

The Role of Beliefs in the Natural Sciences

T0130817

Edited by

Jitse M. van der Meer

The Pascal Centre for Advanced Studies in Faith and Science
Redeemer College, Ancaster, Ontario
and
University Press of America, Inc.
Lanham • New York • London

Copyright © 1996 by
University Press of America, ® Inc.
4720 Boston Way
Lanham, Maryland 20706

3 Henrietta Street
London, WC2E 8LU England

Copublished by arrangement with
The Pascal Centre for Advanced Studies in Faith and Science

Library of Congress Cataloging-in-Publication Data

Facets of faith and science / edited by Jitse M. van der Meer.
        p.   cm.
Includes bibliographical references and index.
Contents: -- v. 2. The role of beliefs in the natural sciences --
v.3. The role of beliefs in mathematics and the natural sciences.
    BL240.2.F33 1996b   261.5'5--dc20   96-41083 CIP

ISBN 0-8191-9990-7 (cloth: alk. ppr.)
ISBN 0-8191-9991-5 (pbk.: alk. ppr.)

The Pascal Centre, established in 1988 by Redeemer College, specializes in studies of the relationship between faith and science from a biblical perspective. The Centre intends to encourage an open-minded discussion of issues in faith and science from this perspective. However, the opinions expressed in this volume are strictly those of the contributors and do not necessarily represent the Pascal Centre or Redeemer College.

The Pascal Centre for Advanced Studies in Faith and Science
Redeemer College
777 Highway 53 East, Ancaster, Ontario
Canada L9K 1J4

Contents

Part 2
Beliefs in the Biological Sciences

PREFACE

The Pascal Centre for Advanced Studies in Faith and Science undertook a five-day research conference in 1992 for scholars with a professional interest in the relationship between science and belief. The objectives were to review the current state of scholarship on the nature of the interactions between the natural sciences and belief, and to identify research that promises new insights. These objectives were pursued along historical and philosophical lines, focusing on the relationship between Christian religious beliefs and the natural sciences. This included issues in theology, history of science, philosophy of science, mathematics, the natural sciences and logic. The result is presented in four volumes under the title *Facets of Faith and Science*, written for the well-educated nonspecialist.

Integration of topics was encouraged in three steps during the conference. First, the main papers and responses provided a broad context for the workshop papers and brought the participants up to date on important questions and answers. Second, specific problems and case studies were discussed in workshops. Workshop chairpersons surveyed the theme of a workshop, and reporters summarized problems, answers and directions for future research. Integration within workshops was encouraged by having papers address the same questions from different angles. To foster exploration and dialogue, the workshop papers were distributed in advance and were discussed without time-consuming presentation. The final integrative step was begun by the concluding panel in which the reporters presented an overview of the workshops. Discussions are not reported because authors had the opportunity to incorporate them into their papers. The conference papers were then transformed into chapters by a peer review process.

The quality of a publication covering the diversity of disciplines encountered in religion and science studies could not possibly be assured without the help of many specialists. I am grateful for the help and encouragement I received from M. Elaine Botha, Roy A. Clouser, Christopher B. Kaiser, Arie Leegwater, Stephen C. Meyer, George L. Murphy, Robert C. Newman, Clark Pinnock, Richard S. Westfall, Stephen J. Wykstra, Delvin Ratzsch, Danie F.M. Strauss, Liba Taub and Uko Zylstra. They have served either as workshop conveners or as consulting editors, or in both capacities. Thanks are due also to the anonymous reviewers and to the authors for the time, effort and patience which they have invested in this work. To Lynda Cockroft, administrative assistant of The Pascal Centre and managing editor of these volumes, I owe a special debt of gratitude, both for her efforts in making the conference a success and for the Herculean task of compiling and preparing for publication the papers included in these volumes. I also thank assistant editors James de Boer, Peter Denton, Erin Fokkes, Theodore Plantinga, Marina van der Meer, Robert Vander Vennen and Alida Van Dijk for their contributions. This project was supported in part by the Social Sciences and Humanities Research Council of Canada.

Jitse M. van der Meer

GENERAL INTRODUCTION

The interactions among religious belief, metaphysical belief, observation, and theory that shape our understanding of the world have been amply documented. Current scholarship on the history of these interactions has distanced itself from past simplifications. What has emerged is a dynamic picture that is approaching the complexity of life itself. This picture has given new impetus to the question about the nature of the interactions between religion and science. How does what we know about God affect what we know about nature, and conversely, how does what we know about nature affect what we know about God? The objectives of these four volumes are to explore how religion and science interact and how deep the relations penetrate into the details of understanding God and nature. These questions are explored from a perspective that integrates historical and philosophical studies. The treatment is not comprehensive. For instance, the role of epistemic values is prominent, but biomedical and environmental ethics are excluded. The role of cognition and in particular of metaphor in mediating between religion and science, as well as the role of the Scriptures in their interaction, are not represented as broadly as the editor had hoped.

Volume 1, Part 1 surveys the historiographical difficulties in establishing the nature of the relations between religion and science. This overview is intended to encourage sensitivity to a diversity of possible misinterpretations of the interaction in historical case studies. Part 2 contains evaluations of integration, conflict and separation perspectives on the relation between religion and science. The objective is to begin an inventory of modes of interaction as a contribution to a more comprehensive understanding of the interaction of religion and science. Volumes 2 and 3 through, in-depth case studies, critically examine the claim that Christian religious beliefs have affected the natural sciences. The purpose is to determine the nature, the extent and the depth of this relation. Volume 2 does this from the perspective of the historic Reformation as it developed into a school of systematic reflection on the meaning of the Christian faith for scholarship also known as the Augustinian, the Kuyperian or the Dutch neo-Calvinist tradition. Introductions to the philosophy of science of three of the fathers of this school—namely, Abraham Kuyper, Herman Bavinck and Herman Dooyeweerd—precede chapters on the role of beliefs in the natural sciences in general, as well as in biology, mathematics and physics. Volume 3 continues with similar contributions from inside and outside of this tradition, covering astronomy, biology, cosmology, physics and systems science. Both Volume 2 (Chapters 4, 12, 13) and

Volume 3 (Chapters 13, 14) discuss hierarchy theory, revealing that there are different conceptions of hierarchy, that the choice among them depends on metaphysical beliefs, and that the failure to distinguish them leads to empirical inadequacy. Volume 2 (Chapter 13) and Volume 3 (Chapter 13, 14) also deal with the relation between metaphysical beliefs and conceptions of organism. Whereas Volumes 2 and 3 explore how religious beliefs inform scientific theories, Volume 4 explores how God acts in nature and history. Part 1 studies how divine action relates to human action and natural processes and how naturalism transformed a view of God's action in nature as transcendent intervention into immanent action through the natural order. Further, Part 1 explores how God's action in nature relates to randomness and to evolutionary interpretations of nature, whether one can identify his immanent action as divine action, and whether God's action in nature can be used as an explanatory strategy in science and as an argument for his existence. It develops philosophical and theological perspectives on the question of how God acts in nature and history. Part 2 presents historical and systematic studies of the hermeneutical questions that arise from the use of the Bible in the interaction of faith and science.

INTRODUCTION TO VOLUME 3
PART 1: BELIEFS IN THE PHYSICAL SCIENCES

In "Astral Piety, Astronomy and Ethics in the Ancient Mediterranean World," Liba Taub explores the interaction among astral religion, ethics and astronomy in Greek antiquity, in particular in Plato and Ptolemy. Whereas astral piety in Near Eastern religions was primarily concerned with worship and salvation, the Greeks studied astronomy, mathematics and philosophy to understand the Good as a guide for living a moral life. This is apparent in the earliest available texts, those of Homer and Hesiod, which reveal that celestial bodies were considered to be created divinities with which one could have a religious relationship. Because celestial phenomena were seen as manifestations of divine action, they were not only explained in terms of divine action, but they were also taken as divine instruction in the moral life. This way of knowing the divine was complemented by knowledge from recollection of the soul's journey among the stars.

Celestial moral instruction could affirm or deny the world. World affirmation in the form of moral guidance for the earthly life and victory over evil was obtained from a study of the stars, including the mathematics and philosophy used for that purpose. World denial was expressed in the use of heavenly knowledge as an escape from the earthly evil into the certainty and stability of the heavenly realm. Taub concludes that the influence of Plato and Ptolemy on later astronomers should be considered, especially with regard to the ethical component which distinguishes their philosophies from other forms of astral religion.

Liba Taub reveals that a similar unity between an earthly and a heavenly realm existed in Chinese culture of the sixteenth and seventeenth century. In "On the Complexity of the Relationship between Astronomy and Religion" she explains that the Chinese used astronomy to organize civic life and to predict the future. This is why the Chinese were interested in the superior Western astronomy introduced by missionaries. The missionaries, on the other hand, practised and promoted Western astronomy to lure the Chinese to Western Christianity. Taub shows how this situation creates some problems for the historiography of religion and science relations in China. In using publications by the missionaries as source material, one must be careful to consider that they interpreted Chinese culture in a Western context, identifying Chinese deities with the Christian God to render the Gospel more acceptable. It may also be artificial to analyze Chinese culture in terms of interaction between religion and science because, as with some traditions of Western Christianity, Chinese traditions do not separate the spiritual and the temporal. Finally, any "Chinese" conceptions of the relation between religion and science may just be the ones they absorbed from their Jesuit instructors.

Bernard Lightman also addresses the cultural and ideological filters through which one receives views on the relation between religion and science. Little is known about how the religious and metaphysical beliefs advertised in many popularizations function in the science of the professional scientist. There can be little doubt, however, that popularization colors practising scientists' views of science and its relation with religion. Lightman explores the dynamics of religious and metaphysical beliefs in the popular science of the Victorian astronomer R.A. Proctor (1837-1888).

Proctor puts the ancient analogy between the heavens and the earth to new use. Projecting the earthly realm which encompassed both nature and society onto the heavens, this analogy led him to the idea of extraterrestrial life. In reverse the analogy suggested a naturalistic social ethics. He became popular because Victorian readers liked to hear the religious implications of his topic, astronomy. Extraterrestrial life has always caught the public interest, and a naturalistic social ethics provided a divine legitimization of an industrial society in upheaval. Lightman argues that Proctor may have read religious and social ideals into nature partly because the Victorian public liked it. And they liked it because these ideals were offered as divine instruction on how to build society, and because they were suitably adorned with the authority and prestige of science. Thus popular science may be an important source of views on the interaction of religion and science held by practising scientists who need to be aware of the political, social, religious and even monetary rationales attached.

John Byl examines the role of metaphysical and religious beliefs in modern cosmology. He argues that since cosmological theories are underdetermined by observations, metaphysical beliefs, religious beliefs and physical background theories, Christians have an opportunity to include biblical revelation and religious beliefs as criteria for theory choice in their cosmology.[1] He exemplifies his thesis by discussing how various attributes of God and the beliefs in the immortality of the soul have been used in the construction of cosmologies. Byl prefers to refrain from speculation beyond what Scripture and observation have to offer and to hold an instrumentalist position with respect to cosmology.

The difficulties of the historiography of faith and science relations are exemplified by two opposite interpretations of Newton's beliefs about divine providence. Richard S. Westfall argues, on the basis of one of Newton's major theological manuscripts, that Newton was a deist. Edward B. Davis, on the other hand, is prepared to argue that Newton believed that God is intimately involved in all phenomena and, therefore, that Newton did not father deism.

Westfall understands Newton as a man deeply concerned about the fate of Christianity in a naturalistic scientific culture. Newton appears to have attempted to stem the tide of naturalism by arguing that nature can be understood only in terms of intelligent design. In this way science could further Christianity by pointing to the existence of a Designer rather than to his absence. Thus, scientific naturalism is the context for what Westfall sees as an influence of science on Newton's religion. Newton's revisionist theology, specifically his Arianism, can also be understood as an attempt to rid Christianity of the irrationalities he feared would contribute to its decline. However, while Westfall acknowledges the influence of Newton's religion on his science, he could see no trace of any such influence emerging from his theology.

Davis, however, argues that Newton's understanding of God's dominion shaped not only the theological perspective but also the content of his science. Divine dominion is revealed in God's action in the world and is identifiable as divine action because it is characterized by purpose and intelligent design. Theology entered his science because the pervasive presence of God made possible a non-Cartesian, voluntarist conception of matter—one that is independent of extension—which rendered action at a distance more plausible than Descartes's action by contact. This voluntarism, Davis argues, renders implausible the typical deist interpretation of Newton endorsing a remote divine clockmaker and the separation of science from religion. Consistent with this thesis, Davis shows that Newton rejected both the clockwork metaphor and the cold mechanical universe upon which it is based. His conception of the world reflects a deep commitment to the constant activity of God's will unencumbered by the rational restrictions that Descartes and Leibniz placed on God—the very sorts of restrictions that later appealed to the deists of the eighteenth century.

Ian Stuart and Tom Settle propose that the undefinable concepts in theories such as time, reality, entropy, causality and order are the loci of relations with metaphysical beliefs because they are open to different interpretations. Such differences usually reveal themselves only when they are mutually inconsistent. They analyze one class of paradoxes and find that they can be avoided by building into physical theory the metaphysical belief in an observer-independent reality. This requires making a categorical distinction between observer-dependent reality (the empirical reality of quantum physics) and observer-independent reality (the primitive reality of classical physics).

For instance, under the Copenhagen criterion for physical reality, the quantum wave function cannot correspond to a physical reality prior to measurement, for there is no interaction at this stage and, hence, we are not dealing with phenomena. Formally, at this stage the wave function obeys the time-symmetric Schrödinger equation. Measurement causes the collapse of the wave function introducing an irreversible time-asymmetry. The collapse is real, for it corresponds to a measurement, but it cannot be obtained from the collapse of something that is not a physical reality. This becomes obvious in the cosmological application of the Heisenberg uncertainty relations. On the Copenhagen view that time is observer-dependent, and that physical theory is valid under time reversal, one cannot adopt the Big Bang theory of the history of the universe because it entails that time is observer-independent. This inconsistency is removed by incorporating in the theory the belief in an observer-independent reality (necessary realism).

The same conflation of observer-independent and observer-dependent reality occurs when entropy is identified with missing information. A physical interpretation of the Boltzmann relation (entropy) is identified with its cognitive interpretation (information), producing an inconsistency between the primitive concepts of order and causality. It is removed by introducing the belief that the universe can have neither complete order nor complete disorder (the cosmological constraint). Stuart and Settle prohibit paradoxes by building into physical theory, not only observer-dependent empirical knowledge, but also beliefs concerning an observer-independent reality. Therefore, the laws and theories of physics do not, in their view, describe reality. They advance what we believe about it in the light of current thought.[2]

Thaddeus Trenn shows how a bias against supernatural events, such as a resurrection from the dead, may have excluded an interpretation of the so-called Shroud of Turin which is consistent with age measurements of the cloth, but which would have allowed the conclusion that it was Jesus' burial cloth. The accepted view, which is based on the carbon-14 method, is that the Shroud originated in the Middle Ages. This date is not incompatible with an earlier date going back to the time of Jesus, Trenn argues, if secondary radiocarbon had been added to the Shroud by an extraordinary supernatural event accompanying the resurrection. This event could have created both the image of a crucified person and the secondary radiocarbon from existing nitrogen in the fibers of the cloth. Such an addition would have "rejuvenated" the Shroud by making it look as if most radiocarbon originally present had not yet had the time to decay. However, cloth treated to remove secondary radiocarbon had the same age as untreated cloth. This has been taken to mean that there was no contamination with secondary radiocarbon, in which case Trenn's hypothesis would have to be rejected. However, Trenn argues, this result could have been due also to a failure of selective chemical removal which requires a difference in chemical environment between primary and secondary radiocarbon. Trenn explains the failure to consider this possibility in terms of an anti-Christian bias against evidence pointing to extraordinary physical events accompanying the resurrection. He also suggests a way to distinguish between radiocarbon contamination originating externally and secondary radiocarbon produced *in situ* from nitrogen by the postulated event.

Davis's test of the so-called Foster-thesis is an exploration of the relation of religion to science in the seventeenth century. The British philosopher Michael Beresford Foster (1903-1959) is known for his work on the influence of Christian theology, especially of the voluntarist variety, on the *content* of early modern natural philosophy. The method of natural science, Foster argued, depends upon presuppositions about nature which depend in turn upon the doctrine of God. For Foster, the connection between Christian theology and the presuppositions of modern science was itself logically necessary, not historically contingent. While Foster's work was unhistorical, it raised interest in two key questions. How did early modern thinkers construe the relation between God and the creation? And what did they think this relation meant for human understanding of the created order? Davis points out that a genuinely historical research program has emerged as a result of Foster's initial work.

Taking this historically sensitive approach, Davis shows how different theologies of creation undergirded different conceptions of scientific knowledge in Galileo, Descartes and Boyle. He uses them to test the Foster-thesis by sketching their understanding of God's relationship to the human mind, of God's relationship to nature, and of the nature of scientific knowledge. Davis agrees with Foster that theological assumptions were closely associated with conceptions of scientific knowledge, but he disagrees with Foster's belief that Christian theology had caused the rise of modern science.

PART 2: BELIEFS IN THE BIOLOGICAL SCIENCES

According to Sytse Strijbos, Ludwig von Bertalanffy's view of life as a dynamical flow of matter and energy through open systems is the foundation of modern systems philosophy and biology. This metaphor of the open system replaced those of the machine (in mechanism) and of the vital force (vitalism). Strijbos argues that each of these metaphors represents a metaphysical belief about the nature of organism and has shaped biological theorizing. An important question is how one evaluates such metaphors. Evaluation uses empirical criteria. For instance, von Bertalanffy's opposition to the machine metaphor of organism stems from his empirical view that a machine is a closed system whereas an organism is an open system. An empirical problem pointed out by Strijbos is that the open system leaves no fundamental boundary between living and nonliving phenomena or between structure and function. In other words, the distinction between open and closed systems does not coincide with that between living and nonliving systems. Not only are there chemical systems that are open systems, but the artificial chemical open system used by von Bertalanffy as a model for organismic open systems is empirically inadequate because its boundaries are static, whereas organisms have dynamic boundaries. The conclusion is that metaphor can mediate between metaphysical belief and biological theory, and that the adequacy of a metaphor can be evaluated empirically.

Paul Nelson argues that Christian religious beliefs have been and still are constitutive in the logic of two arguments for common descent. These are the argument from divine perfection and the argument from divine freedom. Their empirical counterparts are the argument from imperfect design as manifest in vestigial organs and suboptimal design, and the argument from similarity (homology).

According to the argument from divine perfection, if God created the world, we should observe rational and benevolent design. Since this is rarely observed, either God did not create the world, or his methods of creation are inconsistent with his nature. Such presumed weakness of this popular version of the doctrine of creation is then presented as an argument supporting purposeless evolution. Nelson argues that the argument from divine perfection assumes that we know God's plan for the creation of the world and that God's perfection entails the perfection of the world. Such assumptions, he argues, are theologically superficial. Moreover, the concept of perfection is ambiguous.

The argument from divine freedom says that God's freedom should have produced a greater diversity of organic patterns than are observed. Similarities between organisms (homologies) must be explained in terms of common descent on pain of limiting God's freedom if explained in terms of God's action. This argument, Nelson maintains, assumes that God is not free to limit himself to creating organisms according to certain patterns. Add to this Darwin's premise that scientific explanation should be naturalistic, and the patterns of similarity become evidence of common descent. That is, common descent as conclusion requires assumptions about God's action in the world and about scientific explanation. Darwin's theological premises functioned as metaphysical beliefs in evolutionary biology, coloring what was thought possible or reasonable.

Nelson concludes that the use of arguments from divine perfection and freedom in evolutionary reasoning is inconsistent with the principle of methodological materialism. However, rather than keeping theology out of science, he recommends that if science wants to pursue truth, it will have to deal with theological problems. Likewise, if theology wants to speak to all of reality, it will have to deal with scientific experience. Problems arise, not because of the mutuality of the interaction, but because of the superficiality of the theology and the science. Nelson's conclusion depends on how theological premises function in the theory of common descent. What if these premises are part of a methodological theism assumed for the sake of showing the presumed internal inconsistencies of an unsound creationist argument? If they are, Nelson's argument shows that the attempt failed due to poor theology, but not that theology informed the content of Darwinian or neo-Darwinian explanation. This would require proof that the theism functioned epistemologically and not just methodologically. Moreover, proving one particular belief in creation wrong does not entail the truth of purposeless materialistic evolution. The falsity of this dichotomy has been overlooked both by creationists and evolutionists including Darwin and Stephen Gould.

Marvin McDonald studies how religion and science interact in the work of Donald MacKay and Roger Sperry on the interaction of mind and brain. They share a theory of consciousness which is emergentist with a mutual causal determination of mental and neural events. This mutuality implies the efficacy of downward causation by mental events which excludes reductionism from their metaphysical beliefs. It also incorporates functionalism by which they exclude the view of mind as a nonphysical substance ("substance" dualism). However, on other issues they interpret their theory of consciousness differently, possibly on account of metaphysical, methodological and religious beliefs.

For MacKay, the mutuality of mental and neural events makes it possible to address both successfully in the brain sciences. Theoretical knowledge about an agent's mind can be based on an outside observer's interpretation of two forms of data: the agent's personal experience as communicated to the outside observer and the agent's brain function as observed by the outside observer. This outside perspective complements an inside perspective produced by the agent's personal experience of ontological aspects of himself because some of these aspects are unavailable to an outside observer. Since the phenomena of conscious agency cannot be directly observed by the outside observer, the neuroscientist must model conscious agency by obtaining its properties from the data of experience and correlating these properties via information theory with those of brain functioning. This approach maintains the complementarity of the inside and outside perspectives even while the embodiment of mental activities in neural events can be explored by scientists committed to the objectivity of the outside perspective.

The implication of the complementarity of inside and outside perspectives for relations between religion and science is that science can study religion. For MacKay, spiritual activities are embodied in psychological ones, as are the latter in brain processes. This makes possible a psychology of religion using reported features of religious experience in much the same way that brain science uses the properties of conscious agency. However, outside the realm of embodiment, science has no metaphysical or religious implications. MacKay's Christianity did not exclude his methodological versions of determinism and reductionism.

An emergentist theory of consciousness also informs Sperry's views on religion and science. The emergence of different levels of organization led Sperry to a pluralist philosophy of science characterized by a limited reductionism. For instance, he considered the debate on free will versus determinism pointless because for him both terms refer to aspects of downward and upward causation. However, for Sperry, unlike for MacKay, emergence implies that mental events (including religion understood as a system of values) can be explained by science and must be justified by science. Sperry's reduction of religion to values and of knowledge to scientific knowledge also distinguishes his approach from MacKay's.

For the reasons why MacKay and Sperry draw different metaphysical conclusions from the same theory of consciousness, McDonald points to differences in philosophy of science, research experience, theoretical frameworks, metaphysical and methodological assumptions and models of science and religion interactions. Their relative weight in the interaction of religion and science remains to be determined.

David Wilcox proposes a mechanism for God's action in the world in the context of a distinction between primary and secondary causes. He compares the empirical adequacy of four types of hierarchy as potential ontic frameworks for secondary causation. The so-called "control hierarchy" is proposed as the kind that best accounts for organisms and that best describes God's action in the world as we know it from the Bible. This control hierarchy employs the notion of downward causation as a metaphysical category to describe God's action in the world. It also incorporates biblical notions such as creaturely obedience and normative prescription, and has as its condition a designer or ultimate downward (primary) cause.

Strijbos traces the implications of two metaphysical beliefs, the postulates of continuity and discontinuity, for the problem of reductionism. He uncovers a conflict between the continuity and discontinuity postulates in the way Ludwig von Bertalanffy and Herbert Simon use hierarchy theory. Von Bertalanffy reduces the different levels of the hierarchy to their formal-mathematical similarities (isomorphisms) despite his intent to use the concept of hierarchy to counter reductionism. Strijbos claims that the tension also surfaces in Simon's distinction between nature and artifact because he introduces it as a basis for his science of the artificial, but ignores it in his treatment of the human mind as a machine. Strijbos's claim can be questioned, however, because treating the mind as if it is a machine does not obliterate the distinction between the natural and the artificial and is, therefore, not sufficient reason to consider Simon a reductionist. Simon considers himself a metaphysical materialist who practices methodological holism. There is no tension in this position either. The conflict is real, however, in von Bertalanffy, and is due, according to Strijbos, to his failure to differentiate between the hierarchies of "entities" and of "modes of existence." Strijbos suggests the latter hierarchy effectively counters reductionism.

NOTES

1. The suggestion to use the Bible in cosmology introduces hermeneutical questions, some of which are taken up in Part 2 of *Facets of Faith and Science. Volume 4: Interpreting God's Action in the World* edited by J.M. van der Meer (Lanham: The Pascal Centre for Advanced Studies in Faith and Science/University Press of America, 1996).

2. Stuart and Settle's prohibition of paradox prohibits the identification of two Dooyeweerdian aspects of reality, namely the analytical and the physical aspects. Compare R.A. Clouser, "A Sketch of Dooyeweerd's Philosophy of Science" in *Facets of Faith and Science. Volume 2: The Role of Beliefs in Mathematics and the Natural Sciences: An Augustinian Perspective* edited by J.M. van der Meer (Lanham: The Pascal Centre for Advanced Studies in Faith and Science/University Press of America, 1996).

Part 1:
Beliefs in the Physical Sciences

1

Astral Piety, Astronomy and Ethics in the Ancient Mediterranean World[1]

Liba Taub

Ancient Mediterranean spirituality incorporated various forms of astral piety, the veneration of the heavenly bodies as divine or eternal. Evidence of astral piety may be found in diverse literary and artistic works, as well as in some archeological remains (particularly the temples of the cult of Mithras, which often had a curved ceiling decorated with constellations).[2] Astral piety also found expression in the writings of several ancient philosophers.[3]

The veneration of the celestial bodies as divinities was often criticized by Jewish and Christian theologians. Moses warns the people of Israel:

> Nor must you raise your eyes to the heavens and look up to the sun, the moon, and the stars, all the host of heaven, and be led on to bow down to them and worship them; the Lord your God assigned these for the worship of the various peoples under heaven.[4]

A later Jewish writer, responsible for the *Wisdom of Solomon*, also criticizes the veneration of celestial bodies.

> Fire, wind, swift air, the circle of the starry signs, rushing water, or the great lights in heaven that rule the world—these they accounted gods. If

it was through delight in the beauty of these things that men supposed
them gods, they ought to have understood how much better is the Lord
and Master of it all; for it was by the prime author of all beauty that they
were created.[5]

Amongst Christian critics of astral religion, Clement of Alexandria
(second century A.D.) was particularly outspoken:

Some men were deceived from the first about the spectacle of the
heavens. Trusting solely to sight, they gazed at the movements of the
heavenly bodies, and in wonder deified them, giving them the name of
gods from their running motion. Hence they worshipped the sun, as
Indians do, and the moon, as Phrygians do.[6]

As Clement suggests, the Greek tradition of regarding the celestial
bodies as divine was deeply rooted and long lived.

The divine nature of the celestial bodies was emphasized by the
philosopher Plato, who argued that the study of astronomy could play
an important role in ethics. This Platonic notion of an ethical
motivation to study the heavens was adopted by at least one influential
astronomer, Claudius Ptolemy.

THE EARLY GREEK POETS

In the earliest extant Greek texts, those of Homer and Hesiod, there are
important relationships between celestial phenomena and divinity.
Without entering into the problems involved in establishing connections
among myth, religion and personal belief, it can be noted that in the
Homeric poems, the actions and activities of the gods serve to explain
many of the phenomena of the sky. For example, the immortal gods
are described as holding the heaven (*ouranos*).[7] The sun, Helios, is a
god, as is rosy-fingered Dawn. Many of the phenomena of the sky are
either personified as gods or epiphanies of gods; other phenomena are
sent directly by Zeus. It would be a mistake to consider Zeus to be
merely a symbol of nature, a symbol of particular natural phenomena
in the sky. The Homeric poems explicitly characterize Zeus, among
others, as a personification of natural phenomena, but he is not lord of
all. Master of the sky, he is also more than the sky; some aspects of
his life do not involve the sky at all. Nevertheless, the Homeric
understanding of celestial phenomena seems to be intimately connected
to the behavior of the gods. (We will, for the moment, postpone our
consideration of Hesiod's poems.)

However, as the Greek philosophical tradition appropriated the
traditional view of the divinity of the celestial bodies, philosophical
astral piety did not simply entail the veneration of divinity; notably, a
particular form of astral piety was linked to a tradition of teaching that

ethical benefits may be derived from the study of the heavens. Some aspects of the relationships among astral piety, ethical philosophy and astronomy in antiquity are considered here.

First, the Greek philosophical tradition of astral piety must be distinguished from other contemporary practices. The celestial bodies were also regarded as divinities within several Near Eastern religious traditions. While there is evidence that these religions influenced Greek ideas regarding celestial divinity, the nature and extent of this influence is not well known, neither in general terms nor with regard to the possible influence of local religious practice on particular communities, or on individuals. Even though the general subject will not be treated here, several important differences should be pointed out between the Near Eastern and Hellenistic astral religions and the astral piety described in the Greek philosophical, mathematical and astronomical texts. Astral religions are characterized by an emphasis on ritualized practices and a strong interest in salvation and the afterlife. These characteristics are not present in the Greek philosophical tradition.[8] Rather, the philosophical and mathematical texts emphasize the ethical motivation to study the divine celestial motions, stressing that this intellectual endeavor itself is a striving for a noble and disciplined character, which enables the participant to become more similar to the celestial divinities.

Returning to Hesiod, it was in the eighth-century poem *Works and Days* that the first suggestion appears that knowledge of the motions of the celestial bodies may be useful for man, not only for practical purposes (for example, agriculture and navigation), but to help him overcome the evils of the world as well. In fact, the primary purpose of the work as a whole appears to be moral. While "at first sight such a work seems to be a miscellany of myths, technical advice, moral precepts, and folklore maxims without any unifying principle," the poem is composed of four parts which together form a unified whole.[9] The first part of the poem describes the origin and subsequent spread of evil in the world. The second part explains how man may escape these evils through industry, especially in agriculture and trade. A series of maxims useful in everyday life comprises the third section. The final section, which may originally have been a separate work not even written by the same author, states which days of the month are favorable for industry and agriculture. The four parts of the poem are linked by their moral aim, which is to show men how to live in a difficult world. Man may overcome the evil present in this world through industry; astronomical lore allows one to know which days will be particularly favorable for any given task.

PLATO, ASTRAL PIETY AND ETHICS

Many centuries later, this theme was more fully developed in the writings of Plato. Experiencing the disillusionment with the traditional gods and their accompanying morality that many at the turn of the fifth century seemed to feel, Plato sought to establish a religion with visible gods, demonstrating teleological order and an ethics based on the emulation of that order.[10] In the *Timaeus,* the heavenly bodies are gods who, together with the traditional gods of mythology, made the other living things of the universe, including man. However, each human soul has a portion of the immortal soul created by the Demiurge.[11]

It is by means of the celestial gods that ordinary men have several opportunities to gain understanding of the universe. The first opportunity illustrates and depends on Plato's theory of recollection. This theory is first introduced in the *Phaedo* and the *Meno.*[12] According to the account of Timaeus, the Creator thought it proper to create the number of human souls to be equal to the number of stars. Before human souls were given their bodies, each soul was assigned to a star, for a ride through the universe, during which the soul was shown the laws of destiny.[13] This knowledge, acquired by man before his birth, would later have to be recollected.

Socrates's second speech in the *Phaedrus* coincides in many ways with Timaeus's description of the soul's celestial journey before birth and subsequent reincarnations. In Book Ten of the *Republic,* the myth of Er recounts how souls after death (that is, between reincarnations) encounter a model of the astronomical revolutions and harmonies linked to the laws of destiny and then must choose their next life, to be lived in compliance with what has been seen. Once the human soul has acquired its body and begun earthly life, it has other means, also dependent on the celestial bodies, by which to acquire knowledge.[14]

Man has been given the power to perceive the cosmic order for an ethical purpose:

> God invented and gave us sight in order that we, seeing the revolutions of reason in the heaven, might profit by them for the revolutions of our own intelligence, which are kindred to those,...and that learning and sharing in naturally correct reasoning, we might imitate the completely unwandering revolutions of god and stabilize our own wandering motions.[15]

This visibility of the celestial bodies, generated gods, was part of the cosmic plan; the Demiurge made the celestial bodies mostly of fire, so that they might be most bright and fair to see. Unlike the traditional gods of myth who reveal themselves only insofar as they will, the celestial gods are readily seen by all men. This visibility was purposefully ordained by the Creator who, when he ordered the movements of the visible universe, provided a means by which this order could be recognized.[16]

Of the various human sensations, only sight and sound touch the soul.[17] According to Plato, vision is the sensation of motion; man's faculty of vision allows the divinely ordained motions of the heavens to be carried directly to our souls.[18] Timaeus explained that a human being does not perceive objects, but external motions, which, being imparted through sense organs, become internal motions.[19] The sense of hearing also provides access to the divine order, enabling man to perceive the cosmic harmony, the motions of which are akin to the revolutions of our own souls. Indeed, harmony itself "was given by the Muses...as an ally for the inner revolution of the soul which has become discordant, to bring it into order and consonance with itself."[20] When a low sound (a slow motion) overtakes a high sound (a swifter motion), a harmonious sound is perceived which imitates the divine harmony in which, likewise, swift motions are seemingly overtaken by slower motions.[21] These harmonious sounds are most pleasant to man. Once the ordered motions of the universe have been perceived, through sight or hearing, man is in a position to order his own soul similarly and to imitate the divine.

But while every human has several means by which to imitate the eternal cosmic order, Timaeus claims that only those few who are philosophers will be able to possess the fullest measure of immortality which is appropriate to man. In the *Republic*, Socrates voices a similar opinion: "the philosopher, being familiar with the divine order, will himself become orderly and divine as much as is possible for man."[22]

The ultimate goal of philosophy itself is ethical, to acquire knowledge of the Good. According to the *Republic*, the philosopher strives for knowledge of true reality, which cannot be seen and can only be apprehended by thought. Socrates explains in the *Phaedrus* that "Reality exists without shape or color, intangible, the pilot of the soul, visible only to reason; true knowledge is knowledge of it." In the *Symposium,* Socrates claims that once having ascended the ladder of love, "beholding beauty with the eye of the mind, man will be enabled to bring forth, not images of beauty, but realities...and having brought forth and nourished true virtue he will become the friend of God and be immortal, if mortal man may."[23]

According to the program of education for philosophers outlined in Book Seven of the *Republic*, mastery of dialectic and knowledge of reality can only be acquired after preliminary training in music and gymnastics, followed by a course of mathematical studies which includes arithmetic, geometry, stereometry, astronomy and harmonics. The type of astronomy to be studied by future guardians is of a special type and this study will only be undertaken by a few.

In the *Laws,* Plato proposes an educational scheme for the general populace.[24] Although not in training as philosophers, all citizens must pursue a certain amount of astronomical study, for more than the practical value which it may have. For the Athenian, the necessity for study of the heavens was not practical, but religious. He insists that the citizens and their young people learn enough about the heavenly gods to prevent blasphemy of them, and to ensure piety in sacrifice and prayer. (Recalling the relationship between religion and society in the Greek world, such religious activity would be regarded as reflecting the responsibility of the individual to society at large rather than as a type of personal devotion.) Even if one is not capable of attaining knowledge of the heavens, one should at least acquire true opinion.

This religious value which is placed on the study of astronomy is echoed in the *Epinomis* which was probably written by a member of the Academy, if not by Plato himself. Here the Athenian states that the greatest of human virtues is piety, which he claims can be learned through studying astronomy, for it gives man an understanding of "the generation of divine things, the most beautiful and divine of sights which god has enabled man to see."[25]

The study of astronomy plays an important role in Plato's ethical philosophy. Motions of the celestial bodies provide the first step in man's ability to recollect the divine knowledge which he learned before he acquired his body. The appearances of these same motions provide other opportunities for all men to live ordered lives, for any man who has either his sight or hearing intact may perceive the divinely ordered motions of the celestial bodies, and thereby order the motions of his own soul accordingly. Plato's educational program for all citizens requires the study of astronomy, as does the more rigorous scheme proposed for philosopher-rulers. For Plato, it is an ethical necessity for all men to be familiar with the celestial bodies and their motions.

PLATO'S IDEAS AND LATER AUTHORS

As we have seen, several Platonic dialogues contain elements of astral theology and an ethics based on knowledge of celestial divinities. However, his views regarding the ethical role of mathematics, particularly astronomy, failed to attract the interest of his students. For instance, in late antiquity Simplicius credited Plato with having defined the classical problem of astronomy, namely, to account for the apparent, and seemingly complex, movements of the celestial bodies solely in terms of uniform, circular motions. No ethical benefits are included in the statement of the problem of astronomy. While many of Plato's students, including Speusippus (circa 407-339 B.C.) and Xenocrates (head of Academy 339-314 B.C.), did not neglect the study of either mathematics or ethics, they said nothing about the ethical status of mathematics.[26] Even Aristotle, who certainly held the view that the celestial bodies were divinities, did not subscribe to Plato's views regarding the ethical benefits associated with astronomical study. Furthermore, there is no evidence that mathematicians themselves took an interest in Plato's suggestion.

However, in the first century A.D., the Alexandrian Jewish philosopher Philo, following tradition, contrasts the nature of the heavenly bodies with those in the terrestrial region, explaining that

> the sun and moon and the whole heaven stand out in such clear and plain distinctness because everything there remains the same and, regulated by the standards of truth itself, moves in harmonious order and with the grandest of symphonies; while earthly things are brimful of disorder and confusion and in the fullest sense of the words discordant and inharmonious.[27]

Philo agreed with the Platonic teaching that the best thing in the world to contemplate is the heavens, knowledge of which is especially prized, "for just as heaven, being the best and greatest of created things, may be rightly called the king of the world of our senses, so the knowledge of heaven, which the star-gazers and the Chaldeans especially pursue, may be called the queen of sciences."[28] Echoing the Platonic emphasis on the importance of vision with regard to knowledge of the heavens, Philo claims that "the eye is the noblest of the body's members because it contemplates the heaven which is the noblest part of the universe."[29] It was noteworthy to Philo that the patriarch Abraham was principally occupied with the study of the heavens before he encountered God, indicating that such study was worthy of the founder of a new religion.[30]

In the second century A.D., several influential writers expressed a special commitment to Plato's views and particularly pointed to the beneficial aspects of mathematics. In their writings, Theon of Smyrna (circa A.D. 115-40) and the author of the *Didiskalikos* each used the language of the mystery cults of the period in their outlines of Platonic philosophy. In their writings, the various branches of mathematics were described as steps in the purification process, necessary preliminaries to initiation into the greater "mysteries" of philosophy.[31] However, their quasi-religious imagery did not contain the strongly ethical component present in Plato's own writings.

The ethical value of mathematics is discussed briefly by Nicomachus (between A.D. 50 and 150) in his *Introduction to Arithmetic*, a primer of Pythagorean number theory containing brief allusions to Platonic concepts. For Nicomachus, numerical relations illustrate the primacy of the beautiful, definite and intelligible over their opposites. The rational soul orders the irrational, whereby ethical virtues are derived from the resulting equilibrium. The study of arithmetic provides evidence of the order of the universe, for numbers provide cosmic order, as well as orderliness, to human life.[32]

This emphasis on orderliness was central to Pythagorean number theory and also to that branch of mathematics known as harmonics. That music imparts an ethical influence, aiding man in the cultivation of virtue, is an idea present in Plato's writings, echoed also in the writings of the second-century Alexandrian astronomer Ptolemy.[33]

The idea that mathematics in general could impart important ethical benefits to those who pursued its study most likely originated in the teachings of the Pythagoreans, if not in the writings of Plato.[34] That such a doctrine would appeal to mathematicians should not be surprising; what is surprising is the lack of evidence suggesting that ancient mathematicians adopted this point of view. While Theon, the author of the *Didaskalikos*, and Nicomachus all extolled the benefits for man of studying mathematics generally, it should be remembered that these men were not original mathematicians, but rather popularizers writing elementary handbooks on mathematics. The emphasis on the ethical benefits of mathematics thrived within what was largely a didactic tradition.

PTOLEMY AND THE ETHICAL MOTIVATION TO STUDY THE HEAVENS

This didactic tradition reached its zenith in the work of the second-century Alexandrian Claudius Ptolemy, one of the outstanding figures in the history of astronomy. Ptolemy stated in the preface to the *Syntaxis*, sometimes referred to as the *Almagest*, that he was primarily interested in the study of mathematics, which he regarded as the highest form of philosophy. For Ptolemy, mathematics was not merely an intellectual exercise, but rather an ethical endeavor. In this regard, it is likely that Ptolemy was influenced by Platonic philosophy.

Of the three branches of theoretical philosophy—physics, mathematics and theology—Ptolemy accorded the study of mathematics the highest status of all. He made the radical, and certainly non-Aristotelian, statement that both theology and physics should be called conjecture rather than knowledge. Ptolemy reasoned that theology should not be called knowledge because its subject cannot be seen, whereas physics should not be called knowledge because of the instability and lack of clarity of matter itself. So far as Ptolemy was concerned, mathematics represented the only true kind of knowledge.

Furthermore, for Ptolemy, the subject matter of mathematics is fundamentally linked to the subject matter of both physics and theology, for mathematics is an attribute of all things, both those which are mortal and those which are immortal (that is, divine). In mortal things, which are always changing according to their inseparable form, what is mathematical changes together with them; in immortal things it preserves the unchanging form as unchanged. What is mathematical serves an especially important function with regard to the divine realm; Ptolemy stated that it is the essence of mathematics which preserves the unchanging form of those eternal things which have an ethereal nature, namely, the celestial bodies. In order to understand what is eternal and divine, one must also understand that which keeps the divine eternal and unchanging. Mathematics is the surest path to knowledge of that which is divine and eternal.

Yet, for Ptolemy the value of mathematics was not restricted to the realm of theoretical philosophy. Rather, with regard to "virtuous conduct in practical activities and character," mathematics, "above all things, could make men see clearly; from the sameness, order, symmetry and calm contemplated about the divinities making its followers lovers of this divine beauty, accustoming them to, even as it were breathing into them, a similar state of the soul."[35] According to Ptolemy, man desires to make his soul as similar as possible to the divine; mathematics, while not providing certain knowledge of the divine, nevertheless provides the closest approach to the divine for man. As such, mathematics becomes ethically important.

Ptolemy explained that he thought it fitting to order his own activities with a view to a noble and orderly condition, devoting himself to the teaching of the many beautiful theories, especially to those called "mathematical." In order to become virtuous and *kalos*, Ptolemy intended to teach theoretical philosophy, particularly mathematics. He was especially attracted to the study of "the divine and heavenly" and believed that astronomy "alone is devoted to the investigation of the eternally unchanging." Mathematics (of which astronomy is a type)

is the best science to help theology along its way, since it is the only one which can make a good guess at that activity which is unmoved and separated; it is familiar with the attributes of those beings which are on the one hand perceptible, moving and being moved, but on the other hand eternal and unchanging.[36]

Therefore, by concentrating on astronomy, Ptolemy was dedicating himself to the study of the eternally unchanging, divine and heavenly bodies.

For Ptolemy, and many other ancient writers, the celestial bodies were divine. Furthermore, for him the study of the celestial motions was an ethical endeavor.[37] He was confident that the study of the heavens could result in positive ethical changes for man. By studying the celestial motions and emulating them, insofar as possible, man would become more like the divinities.

The conviction that the study and teaching of astronomy would have ethical consequences clearly motivates Ptolemy in the *Syntaxis*. This same ethical theme recurs in his later work as well. At the beginning of the *Tetrabiblos*, Ptolemy is at pains to argue in favor of the value of astrological prediction. Ptolemy distinguishes between the two types of astronomy: one predicts the motions of celestial bodies and is quite accurate; the other, which we call astrology, predicts events on Earth, and is less accurate, primarily because of the nature of earthly things themselves. He counters the hypothetical argument that foreknowledge of events is superfluous by stating that

> we should consider that even with events that will necessarily take place their unexpectedness is very apt to cause excessive panic and delirious joy, while foreknowledge accustoms and calms the soul by experience of distant events as though they were present, and prepares it to greet with calm and steadiness whatever comes.[38]

Ptolemy regards astronomy, be it the type found in the *Syntaxis* or that of the *Tetrabiblos*, as providing men with a way to order their lives and achieve peace and tranquillity. Furthermore, everyone has access to the benefits of astronomy, either through first-hand experience and knowledge, or by second-hand means from a book or astronomer.

In his search for inner peace, Ptolemy shares the ethical goals of many Hellenistic philosophers, yet he employs different means to achieve those goals. His reverence for the celestial bodies as divinities was also deeply rooted in the Greek philosophical tradition. While his emphasis on astronomy as a key to ethics was unusual, this point of view was not entirely unique in the history of Greek philosophy, having been advocated by Plato many centuries earlier.

Ptolemy proclaimed his own place within this educational heritage by pointing to his role as a teacher. Strikingly, Ptolemy, as had Plato, admitted different ways in which an individual could derive the ethical benefits of astronomy, for example, by studying, by teaching and by making original progress in theories.

Of course, Ptolemy was no mere advocate of the study of mathematics, but one of the most influential contributors in the history of astronomy. It is all the more striking that it is in Ptolemy's astronomical writings, which represent a culmination of Greek astronomy, that we similarly find the culmination of a rather neglected form of Platonic ethical theory, with its special emphasis on astronomy. Ptolemaic astronomy had enormous influence for many centuries to come, and the works of later writers indicate many links between astronomy and religion.[39] In conclusion, while the later history of astral piety is too broad to be continued here, the philosophical influence of Plato and Ptolemy on later astronomers should be considered, with particular attention paid to the ethical component which distinguishes their philosophies from other forms of astral religion.

NOTES

1. Parts of this paper, presented at the First International Pascal Centre Conference on Science and Belief in August 1992, were incorporated into my book, *Ptolemy's Universe: The Natural Philosophical and Ethical Foundations of Ptolemy's Astronomy* (Chicago: Open Court, 1993).

2. There is a vast literature on the subject. For various points of view, see D. Ulansey, *The Origins of the Mithraic Mysteries: Cosmology and Salvation in the Ancient World* (New York: Oxford University Press, 1989); J.D. North, "Astronomical Symbolism in the Mithraic Religion," *Centaurus* 33 (1990): 115-48; N.M. Swerdlow, "On the Cosmical Mysteries of Mithras" (review of Ulansey's book), *Classical Philology* 86 (1991): 48-63.

3. While many of the sources considered here may be regarded as "literary," I do not intend to offer a literary interpretation or rhetorical analysis of the material. Questions concerning the genre, rhetorical structure and audiences of the works discussed here would yield useful insights into the philosophical content, but in most instances I leave those tasks to others.

4. Deuteronomy 4:19. *New English Bible with Apocrypha* (New York: Oxford University Press, 1972). All further biblical references are to this version.

5. *Wisdom of Solomon* 13:2-3.

6. Clement of Alexandria *Protepticus (The Exhortation to the Greeks)* 2.22. The etymological linking of the word for gods (*theoi*) with the word "to run" (*thein*) comes from Plato, *Cratylus* 397 c-d, and is repeated elsewhere, including Aristotle's writings.

7. *Odyssey* 1.67.

8. It is not my purpose here to consider generally the similarities and differences between religion and philosophy, nor would I suggest that any "shift" took place from religious faith to ethical behavior.

9. H.E. White, "Preface" to *Hesiod: The Homeric Hymns and Homerica* (New York: G.P. Putnam's Sons, 1982; reprint of 1914 ed.), xviii.

10. F. Solmsen, *Plato's Theology* (Ithaca: Cornell University Press, 1942), chaps. 1-3.

11. Plato, *Timaeus*, 39e-41a; 41c-d.

12. Plato, *Phaedo*, 73c-76b; Plato, *Meno*, 82b-86b.

13. Plato, *Timaeus*, 41d-e.

14. Plato, *Phaedrus*, 246a-250a; Plato, *Republic*, 614b-621c.

15. Plato, *Timaeus*, 47b-c.

16. Plato, *Timaeus*, 40a-41a.

17. I am indebted to Richard Baker for pointing out that the sensations of sight and hearing alone touch the soul; see Plato, *Timaeus*, 67a7-b7.

18. Plato, *Timaeus*, 47b-c.

19. Plato, *Timaeus*, 45b-d.

20. Plato, *Timaeus*, 47d.

21. Plato, *Timaeus*, 47c-e, 80a-b, 39a-b.

22. Plato, *Timaeus*, 90b-c; Plato, *Republic*, 500d.

23. Plato, *Republic*, 490a-b, 511c-d; Plato, *Phaedrus*, 247c; Plato, *Symposium* 212a.

24. Plato, *Laws*, 820e-821e.

25. [Plato] *Epinomis*, 991b. The *Epinomis* may have been written by Philip of Opus. See L. Tarán, *Academica: Plato, Philip of Opus, and the Pseudo-Platonic "Epinomis"* (Philadelphia: American Philosophical Society, 1975) and A.E. Taylor, "Plato and the Authorship of the 'Epinomis,'" *Proceedings of the British Academy* 15 (1929): 235-317. Jean Pépin, in an article on ancient religious attitudes towards the universe as a whole, briefly discussed the religious aspects of the *Epinomis* in "Cosmic Piety," in *Classical Mediterranean Spirituality: Egyptian, Greek, Roman*, edited by A.H. Armstrong (New York: Crossroad, 1986), 408-435, especially 411-13.

26. Yet, admittedly, not much is known about either of these figures. For Speusippus, fragments and testimonia remain; for Xenocrates, we have only scanty evidence regarding his philosophical writings.

27. Philo, *On Joseph*, 145.

28. Philo, *On Mating with the Preliminary Studies*, 50.

29. Philo, *On the Special Laws*, 3.202.

30. Philo, *On the Cherubim*, 4.

31. Theon of Smyrna, *Philosophi Platonici Expositio rerum mathematicarum ad legendum platonem utilium*, edited by E. Hiller (Leipzig: Teubner, 1878), page 1, lines 1 ff; page 14 lines 18 ff. Alcinous, *Didaskalikos*, in *Platonis Dialogi secundum thrasylli tetralogias dispositi*, edited by C.F. Hermann, 6 vols. in 3 (Leipzig: Teubner, 1859), 28, 182, lines 7 ff. The author of the *Didaskalikos* is difficult to identify; see J. Whittaker, "Platonic Philosophy in the Early Empire," *Aufstieg und Niedergang der Römischen Welt* (1987), Teil II, Band 36, 1:81-123.

32. Nicomachus, *Introduction to Arithmetic*, 1.14.2; 1.23.4.

33. Plato, *Republic*, 522a; Plato, *Timaeus*, 47c-e, 80a-b. Ptolemy, *Harmonics*, 1.2; 3.4 ff.

34. See W. Burkert, *Lore and Science in Ancient Pythagoreanism* (Cambridge: Harvard University Press, 1972), who argued that much of "Pythagoreanism" was intellectual propaganda disseminated by Plato and his followers in order to establish a philosophical heritage for themselves.

35. Ptolemy, *Syntaxis*, 1.1.

36. Ptolemy, *Syntaxis*, 1.1.

37. Ptolemy, *Syntaxis*, 9.2; 4.9.

38. Ptolemy, *Tetrabiblos*, 1.3.

39. As J.M. Lattis has shown, a Christian version of Ptolemaic cosmology found particularly clear expression in the writings of Christoph Clavius (J.M. Lattis, *Between Copernicus and Galileo: Christoph Clavius and the Collapse of Ptolemaic Cosmology* [Chicago: University of Chicago Press, 1994]).

2

On the Complexity of the Relationship between Astronomy and Religion: Jesuit Missionary-Astronomers in China in the Sixteenth and Seventeenth Centuries

Liba Taub

The interactions between astronomy, cosmology, theology and religion during the early modern period have been the subject of much study and debate, but attention has been focused almost exclusively on the Galileo controversy. During the same period there were other examples of the interaction between science and religion which should also be considered if we are to enhance our understanding of the relationships between science and religion. My purpose is to raise interest in another important episode.

During the sixteenth and seventeenth centuries, various European missionaries working overseas attempted to win converts to Christianity. The approach of one group, the Jesuit missionaries in China, was unusual, for they sought to interest the Chinese first in Western science (astronomy in particular) and then, after convincing them of the superiority of Western science, to convert them to Catholicism.

The story of the Jesuit experience in China is long and complicated; we find a complex relationship between astronomy, Confucianism and Catholicism, a surprising link between Christianity and astronomy, and abundant evidence of Jesuit commitment to the study and teaching of astronomy. Here we will only examine a brief period and will confine our consideration primarily to the question: "How did religion interact with astronomy?"

There are some difficulties involved in addressing this question, because of the problems involved in defining religion. John Brooke argues against the adoption of too restrictive a definition.[1] Without attempting to define religion himself, Jacques Gernet warns that "despite the importance of religious phenomena in China, as elsewhere, in China the place and functions of religion are different" from the West. As he suggests, the social-political systems within Europe and China were vastly different; more importantly, there were profound differences in mental categories.[2]

Although there had been prior attempts in the sixteenth century to promulgate Christianity in China, the Catholic missions there had their beginning in the last two decades of the 1580s, with the work of the Jesuit Matteo Ricci (1552-1610). Settling in Peking in 1601, he remained there under the protection of the Wan-li Emperor until his death. A student of the mathematician Christoph Clavius at the Collegio Romano, he mastered the Chinese language, adopted the dress of a Chinese scholar and disseminated Western science through lectures and books in Chinese, maps and scientific instruments. It was his gift of clocks which convinced the emperor that he was a man worth protecting; Michael Ryan suggests that it was the enthusiastic response to the gift which suggested to the Jesuits that science and technology held great appeal for the Chinese.[3]

What is particularly interesting for us is that Ricci hoped to convert the Chinese to Catholicism by demonstrating to them the superiority of Western (Catholic) science. But why should the Chinese be interested in Western science? First, let us consider Chinese interest in astronomy. Astronomy played a central role in the traditional Chinese state, having political, social, ritual and metaphysical significance. Furthermore, the emperor was regarded as the mediator between heaven and earth; this role was imbued with metaphysical significance.

The first duty of the emperor was to establish the calendar to indicate the seasons and the relationship of the civil year to the astronomical year. The central importance of the calendar to Chinese public and private life cannot be underestimated; all activities, from the most mundane to those of the greatest importance were governed by the calendar. The establishment and distribution of the calendar, with its listing of auspicious days, was a primary responsibility of the state.

The calendar also had a religious purpose in that it demonstrated the harmony between heaven and earth. The various possible correspondences between humanity and the cosmos, microcosm and macrocosm are fundamental to Chinese thought. The "whole of humanity, from self to family to state to world, was part of a harmonious cosmic order, overseen by *t'ien* [heaven]."[4] The term "correlative cosmology" is widely used by scholars, following Joseph Needham,[5] to describe the doctrine of the "Interrelation of Heaven and Man" (*T'ien-jen chih chi*); Benjamin Schwartz suggests that the term "correlative anthropocosmology" might actually be more appropriate. He describes the correlative cosmology as being based on the idea of the existence of homologies between human and natural phenomena, which can be used as "a means of controlling human civilization as well as human individual life by 'aligning' them with the cycles, rhythms, and patterns of the natural realm." In Schwartz's view, such a cosmology would normally be considered part of religious thought in the West.[6]

Contrasting the quite different worldviews of China and the West, Frederick Mote suggests that:

> The all-enfolding harmony of impersonal cosmic function can be seen to serve analogous, yet qualitatively different, ends from those provided by cosmologies oriented toward a supreme power that knowingly directs the cosmos. The Chinese world view kept man's attention on life here and now and made Chinese thinkers responsible for ordaining the forms and patterns of that life. The ritualized society of China can be adequately explained only in terms of its own cosmology; the relationship of the one to the other is direct and primary.[7]

Indeed, there is ample evidence to support this perception.

Within the Chinese traditions there is no separation between the spiritual and the temporal.[8] Henderson points particularly to the importance of the correspondence of "aspects of the cosmos, especially the heavens, with the dynastic state and the imperial bureaucracy." Associations between polity and cosmos, humanity and nature had great impact on intellectual pursuits, including the various "sciences and pseudosciences, particularly medicine, alchemy, astronomy, astrology, geomancy and the divinatory arts in general." He notes that in the post-Han period (after 220), even as mathematical astronomers were adopting a view that their astronomical systems should be regarded as artificial accounts of nature, they did not discard their commitment to correlative cosmology, as evidenced in the continuing interest in numerology tied to astronomical and calendrical work. The cosmology of the Ming period (1368-1644) tended to be quite conservative, reaffirming and even extending the cosmological correlations of earlier periods.[9]

The establishment of a new dynasty was marked by the issuing of a new calendar, acknowledging the power and responsibility of the new emperor. Because the imperial title "Son of Heaven" signified the emperor's role as mediator between heaven and earth, problems with the calendar accordingly had both religious and political significance. Failure on the part of the emperor to establish a reliable calendar had grave political, religious and social ramifications, calling into question his celestial mandate and divine right to rule. As Richard J. Smith argues, if "individual Chinese were preoccupied with fate, the Ch'ing government was positively obsessive about it," and with good reason. He explains that:

> imperial legitimacy was viewed, in large measure, as a matter of predicting the future correctly—whether in terms of anticipating events such as eclipses, interpreting portents, or selecting auspicious days for conducting all-important state rituals. Because failure to perform these functions satisfactorily called into question the regime's right to rule by virtue of the heavenly mandate, activities such as regulating the calendar (*chih-li*) and fixing the time (*shou-shih* or *shih-ling*) assumed special importance to every dynasty.[10]

With so much at stake, the calendar was an important political tool and a powerful symbol for the emperor.

But the emperor, even as mediator between heaven and earth, was not expected to determine the calendar on his own; he was not regarded as all-knowing. The emperor, to aid in the establishment of the calendar, organized the Astronomical Bureau as a special government department, whose duties included not only calendrical, but also observational work, for celestial phenomena were regarded as portents. The Astronomical Bureau delivered the new calendar to the emperor each year. The calendar, which included auspicious dates for even very common activities (visiting friends or the barber), was then forwarded to the Ministry of Rites for promulgation. Since so much depended on the reliability of the calendar, the emperor's choice of experts to develop the calendar was crucial.

Most of the astronomical knowledge and techniques used in the period of the Ming dynasty were those of Kuo Shou-ching (1231-1316); at times they proved to be inadequate. During the sixteenth century a crisis developed. In the years 1517 and 1518, discrepancies were already found in the calendar for prediction of solar eclipses, and there were instances when members of the Astronomical Bureau had their salaries suspended because of wrong calculation of lunar eclipses.

It was one of the founding members of the Jesuit order, St. Francis Xavier (1506-1552), who recognized the possible usefulness of European astronomy to aid in Chinese calendrical work. As Xavier saw it, once the emperor converted to Christianity, the Chinese people, and the Japanese, would follow his example; Christianity would soon spread throughout Asia. The Emperor of China was the key to the Christian conquest of the East.

By the time of Ricci's arrival in China in 1583, the Chinese were quite dissatisfied with their calendar. Ricci was sensitive to the importance of a workable calendar; his former teacher, Father Christoph Clavius, with whom he corresponded affectionately throughout his life, had been instrumental in the Gregorian reform of the Julian calendar in Europe, adopted in 1582. Ricci shrewdly realized that if he could help the Chinese with their calendar, he would win their trust and approval, and possibly the conversion of the emperor himself. He too regarded the conversion of the emperor as the primary goal of his missionary work, believing that if he were successful the entire Chinese population would follow the emperor's example.

The Chinese, already familiar with Ricci's work on geography, had figured out that Ricci might be able to help them with their calendar troubles. In 1592 or 1593 the President of the Board of Rites at Nanking invited Ricci to help. By the late 1590s Ricci had developed a methodology of conversion based on imparting knowledge of Western science. He was successful in converting some important Chinese scholars, by beginning with the serious study of Western science, and following with indoctrination in Christianity. Strikingly, many of his converts were committed to the link between Western science and Catholicism. One of his students, Paul Hsü Kuang-ch'i, decided to collaborate with Ricci and

> translate one of our scientific books to show the scholars of this kingdom with what diligence ours carry on their researches and on what solid foundations they establish their proofs; through that, they come to understand that in matters of sacred religion, it was not lightly that we had decided to take the side we did.[11]

Ricci died in 1610 and the emperor donated land for the construction of his tomb, as well as for a residence and chapel for the Jesuits, thereby granting protection to the Catholic missions in China. Ricci's journals were brought back to Europe, where, being the first carefully written accounts of Chinese civilization to reach the West, they went through editions in Latin, French, Spanish, German and Italian. One reader, the young Jesuit Adam Schall, found his vocation in Ricci's

Figure 2.1: Portrait of Adam Schall in *China Monumentis* by Kircher, 1667. Courtesy of the Adler Planetarium & Astronomy Museum, Chicago, Illinois.

writings; he applied and was granted permission to serve in China. Ricci had described the secret of his success as a missionary:

> In order that the appearance of a new religion might not arouse suspicion among the Chinese people, the Fathers did not speak openly about religious matters when they began to appear in public. What time was left to them, after paying their respects and civil compliments and courteously receiving their visitors, was spent in studying the language of the country, the methods of writing and the customs of the people. They did, however, endeavor to teach this pagan people in a more direct way, namely, by virtue of their example and by the sanctity of their lives.[12]

Following Ricci, Schall applied himself to astronomy and the problem of the calendar.

Schall began his astronomical career in China by correctly predicting the eclipse of October 8, 1623. In 1629 Hsü Kuang-ch'i, a former student of Ricci and a convert to Christianity, was made head of the Calendrical Department of the Board of Rites. He engaged the Jesuit Terrentius (Johann Schreck) to join his staff. Terrentius, along with Galileo, had been a member of the Lincean Academy and was a regular correspondent of Kepler. When Terrentius died in 1630, Schall was appointed to take his place and continued both his calendrical and missionary work. In 1644 Schall was appointed as director of the Bureau of Astronomy. Taking advantage of his new authority, his astronomical publications often included attempts to teach Christian morality.

Following the contemporary Jesuit policy of accommodation and adaptation to native cultures, Schall did little to upset traditional beliefs.[13] Adopting the advice of Xavier and the example of Ricci, he believed that his primary goal must be the conversion of the emperor. In order to gain access to the emperor he realized, as had Ricci, that he must have the respect of important Chinese scholars and officials. Like Ricci, he mastered Chinese and the Confucian classics, and he adopted the robes of a Confucian scholar. Schall adopted the Jesuit policy of accommodation, adapting his methods to the context, as Jonathan Spence explains:

> To translate the word "God" he used accepted terms for "Deity" and "Lord of Heaven" culled from the Chinese literary heritage, assuring his European critics that there were theistic elements in the Confucian canon that made such usage legitimate. Furthermore, he accepted the idea that the rites which the Chinese performed in honor of their ancestors or of Confucius had purely civil significance, and that Chinese converts to Christianity might continue to practice such rites without being condemned as heretical.[14]

He was joined in his work by another Jesuit, Ferdinand Verbiest (1623-1688), who was summoned from Shensi province to Peking by the emperor in 1660.

Most of the accounts of the Jesuits in China reflect one of two primary concerns: a particular interest in Jesuit history or the desire to study East-West relations. Few modern scholars have been concerned with the relationship between Catholic theology and astronomy as they were intertwined in the Jesuit missionary effort in China. Yet contemporary Catholics, including other Jesuits, but particularly members of "rival" orders, were critical of Schall's approach, arguing that it had serious theological implications. Further, as the Dominicans and Franciscans came to China and attacked the lavish life-style of the Jesuits, Jesuit practices were condemned by a series of Church officials, including the Pope himself, from 1635-1655. (The Pope withdrew his prohibition in 1656, which furthered angered the Dominicans.) A Franciscan voices the criticism:

> The Jesuits have selected purely human means to spread the faith, in direct contradiction to the means used by the Apostles, recommended by Christ Our Lord, and employed by all those who subsequently worked to spread the Kingdom of God—with the exception of said Jesuit Fathers.[15]

Chinese astronomers also attacked Schall, leading to the proscription of Catholicism, the closing of churches and the banishment of missionaries to Macao and Canton. He bore the brunt of various jokes among the missionaries themselves: "Father Ricci got us into China with his mathematics, and Father Schall got us out with his."[16]

Schall died under a great cloud in 1666. He and Verbiest had both been accused of astronomical errors, among other crimes such as spying, and had suffered punishment.[17] Verbiest remained under house arrest until 1669, when he was able to convince the new K'ang-hsi emperor of his astronomical usefulness and was appointed director of the Astronomical Bureau. Having devoted much of his time to calendrical work and the building of observatory instruments, he nevertheless always hoped to convert the emperor, who studied mathematics and astronomy with him. However, in spite of his interest in astronomy, the emperor never did convert to Christianity.

Xavier's dream to convert the Chinese to Christianity was made reality through the missionary work of Matteo Ricci and his followers. That the Jesuit proselytizing effort was linked to the promulgation of Western astronomy was quite remarkable. Ricci's plan to first win converts to European astronomy and then to convince those scientific converts to adopt European religion depended, to a large extent, on his own understanding of the crucial nexus between astronomy and metaphysical beliefs in China.

The story of the relationship between astronomy and religion within the context of the work of the Jesuit missionaries in China can only be sketched broadly here, but I hope that I have given some indication of the richness of the material to be studied.[18] While such an overview can only hint at the complexity of the interactions between religion and astronomy contained in this celebrated episode, it provides a cautionary message for those who would suggest that the "moral" of the story is a simple one. As John Brooke, in his book *Science and Religion: Some Historical Perspectives*, notes, much of the literature devoted to the study of the relationship between science and religion "is suspect because of the thinly veiled apologetic intentions; much is vitiated by an insensitivity to the richness of past debates that historical analysis alone can remedy."[19] He points to several landmark interpretations and their authors. Among these, one theme in particular has often captured the public imagination: the idea that science and religion are at war. Celebrated presentations of this point of view include those of J.W. Draper and A.D. White. Historians have debated the suggestion that the rise of modern science owed an important debt to Protestant Christianity—the so-called Protestant "spur" to science; much of this debate goes back to the work of Robert K. Merton.[20] Brooke acknowledges the value these studies have had, while arguing that such attempts to define in simple terms a single relationship between science and religion are destined to be ultimately unsatisfying because they fail to recognize the richness and complexity of history.

For example, discussions of the relationship between astronomy and religion during the early modern period have tended to be preoccupied with the Galileo episode. Clearly, there were other important interactions between science and religion, even between astronomy and the Catholic Church, taking place at the same time. These should be examined together, in order to yield a fuller picture. Further, those who have written about the Jesuit experience in China have seldom considered the relationship between astronomy and religion itself, telling instead the stories of individual figures, or concentrating on the Rites Controversy or on East-West relations.

Regarding the relationship between science and religion during the period of the Scientific Revolution, Brooke makes two points which seem particularly relevant to understanding the mission of the Jesuit astronomers. He suggests that during the period of the Scientific Revolution, science and religion became differentiated, but not separated. Furthermore, claims regarding the relationship between

Protestant Christianity and the development of modern science should be scrutinized carefully.

The question must be raised as to the degree of separation between astronomy and religion in the minds of the Jesuit missionary astronomers. While this topic deserves detailed investigation, it should be mentioned that Ricci deemed it perfectly appropriate to provide an explanation of Ptolemaic astronomy in his introduction to Christianity written for Chinese scholars, *The True Meaning of the Lord of Heaven* (*(T'ien-chu Shih-i)*. Further, in this work he indicates that "*T'ien*" ("Heaven") and "*Shang-ti*" ("Emperor Above") are synonymous with the Christian God.[21] And the commitment to the study of the heavens, not only to Heaven, must certainly also be acknowledged. Schall explained that "There is nothing surprising in the fact that the Jesuits apply all their energies to the reform of astronomy, since it is a purely scientific occupation, not at all out of place for religious men." Of course this scientific work was presumably for the sake of the conversion of the Chinese. Late in life, Schall explains:

> I thought to myself of all the work that for over twenty years I had given to the reform of astronomy, and I wished that it should not be lost. For it was to astronomy that I was indebted and I hoped for similar results for my successors. At the same time I saw to my great sorrow that this work was held in contempt by those who above all should have encouraged it.[22]

Perhaps he was referring to both the Chinese astronomers and to other Christians as well. Further study will be necessary to understand both Ricci's and Schall's points of view. Further light should be shed by the examination of the contexts of the post-Reformation Church and of the Rites Controversy, during which the Jesuits were accused of laxity in allowing converts to continue traditional rituals associated with ancestor cults.

Furthermore, claims that Protestant Christianity contributed to the development of modern science must be qualified, as Brooke warns, and the context must be specified. The introduction of European science (and especially astronomy) to China was, to an important extent, due to the deliberate efforts of Catholic missionaries. Indeed, the use of science in the proselytizing work of the Jesuit missionaries provided an inspiration for at least one Protestant, Gottfried Wilhelm Leibniz (1646-1716), who also hoped to use science to convert the Chinese to Christianity.[23]

Finally, the reaction of the Chinese to Jesuit astronomy and religion must be considered. Gernet assesses the success of the Jesuits rather negatively; indeed, he criticizes their lack of understanding of the fundamental differences between the worldviews of the two civilizations. He argues that Ricci applied mental frameworks to Chinese thought which were completely alien and, inevitably, committed gross errors of interpretation. For example,

> The classical formulae, "respect" and "fear Heaven," really meant something quite different from the sense given them by Ricci and by many other missionaries after him, who were led on by the mirage of a "natural religion" or the idea of an ancient transmission of the message of the Bible to the Chinese. These formulae did not refer to a single, all-powerful God, the creator of heaven and earth, but instead evoked the ideas of submission to destiny, a religious respect for rituals, and serious and sincere conduct.[24]

Gernet suggests that the Jesuits would have done well to concentrate more fully on their study of Chinese culture.

However, Gernet points to the strong tendency to understand alien ideas within the framework of one's own culture, noting that:

> Just as some Jesuits, called "figurists," believed that in the Chinese Classics they had discovered a whole complex of Christian symbols and transparent allusions to Christianity, similarly, the [anonymous] author of [a] little work [written at the end of the Ming period] interprets what he thinks he knows of the [Jesuit teaching of the] doctrine of the Master of Heaven by means of symbols taken from the *Book of Changes*. It is all very Chinese and not very Christian. In this test Heaven appears not as a way of referring to a personal, creator God, but as the anonymous power whose continuous action ensures the alternations and equilibrium of nature.[25]

In spite of these differences of outlook and understanding, Gernet claims that "for most of the Chinese at the end of the Ming period, Jesuit teaching formed a unified whole: calculations relating to the calendar, astronomy, 'respect for Heaven' and morality—for them, all these things went together." Thus, the term "heavenly studies" (*T'ien-hsüeh*) was generally applied to the teachings of the missionaries.[26] This unified view of the Jesuit teaching in China would no doubt have appealed greatly to the Jesuits themselves, who suggested that there was a fundamental link between their astronomical and religious work.

The hopes of the Jesuit missionary-astronomers are articulated by Verbiest, who writes: "It was a star that long ago led the Three Kings to adore the True God. In the same way the science of the stars will lead the rulers of the Orient, little by little, to know and to adore their Lord."[27] He proudly notes that "astronomy introduced and established

Christianity in China and with good cause she leans on the arm of Astronomy."[28] The answer to the question "how has religion interacted with astronomy?" must surely be "in complex ways, which deserve to be investigated further."

NOTES

1. J.H. Brooke, *Science and Religion: Some Historical Perspectives* (Cambridge: Cambridge University Press, 1991), 6-11.

2. J. Gernet, *China and the Christian Impact: A Conflict of Cultures*, translated by J. Lloyd (Cambridge: Cambridge University Press, 1985), 247; translation of *Chine et christianisme* (Paris: Editions Gallimard, 1982), 3.

3. M. Ryan, "The Diffusion of Science and the Conversion of the Gentiles in the Seventeenth Century," *In the Presence of the Past: Essays in Honor of Frank Manuel*, edited by R.T. Bienvenu and M. Feingold (Dordrecht: Kluwer, 1991; Archives Internationales d'Histoire des Idées 118), 9-40.

4. D.W.Y. Kwok, *"Ho* and *T'ung,"* in *Cosmology, Ontology, and Human Efficacy: Essays in Chinese Thought*, edited by R.J. Smith and D.W.Y. Kwok (Honolulu: University of Hawaii Press, 1993), 1-9, especially 7-8.

5. J. Needham, *Science and Civilisation in China*, 6 vols. (Cambridge: Cambridge University Press, 1956-91), 2:279-93.

6. B.I. Schwartz, *The World of Thought in Ancient China* (Cambridge: Belknap Press of Harvard University Press, 1985), 350; see also chap. 9, "Correlative Cosmology: The 'School of *Yin* and *Yang*,'" 350-383; quoted passages on 355.

7. F.W. Mote, *Intellectual Foundations of China* (New York: Alfred A. Knopf, 1971), 26-27.

8. Gernet, 4.

9. J.B. Henderson, *The Development and Decline of Chinese Cosmology* (New York: Columbia University Press, 1984), 5, 46, 114-15, 132-36.

10. R.J. Smith, "Divination in Ch'ing Dynasty China," in *Cosmology, Ontology, and Human Efficacy: Essays in Chinese Thought*, edited by R.J. Smith and D.W.Y. Kwok (Honolulu: University of Hawaii Press, 1993), 141-78, especially 155-56.

11. Translation in H. Bernard, *Matteo Ricci's Scientific Contribution to China*, translated by E.C. Werner (Peiping: Henri Vetch, 1935), 67. Original in M. Ricci, *Opere storiche del P. Matteo Ricci S.J.*, 2 vols., edited by P.T. Venturi, (Macerata: Filippo Giorgetti, n.d.), 1:500:

di tradurre qualche nostro libro di scientie naturali per mostrare ai letterati di questo regno con quanta diligentia i nostri investigano le cose e con quanto begli fondamenti le affermano e provano; da dove verrebbono a intendere che nelle cose della nostra santa religione non si erano leggiermente mossi a seguirla.

12. [M. Ricci], *China in the Sixteenth Century: The Journals of Matthew Ricci, 1583-1610*, translated by L.J. Gallagher, (New York: Random House, 1953), 154.

13. On the method of accommodation, see J. Bettray, *Die Akkommodationsmethode des P. Matteo Ricci S.J. in China, Analecta Gregoriana* (Rome: Gregorian University, 1955), 76, sect. B, n. 1.

14. J.D. Spence, *The China Helpers: Western Advisers in China 1620-1960* (London: The Bodley Head, 1969), 14. This latter interpretation resulted in the so-called "Rites Controversy," about which there is an extensive literature; see, for example, J.S. Cummins, *A Question of Rites: Friar Domingo Navarrete and the Jesuits in China* (Aldershot: Scolar Press, 1993).

15. Translation in Spence, *China Helpers*, 21. Original in A. van den Wyngaert, ed., *Relationes et Epistolas Fratrum Minorum Saeculi XVII, Sinica Franciscana*, 3 vols. (Florence: Collegium S. Bonaventure, 1936), 3:90:

patres societatis in illis orientis missionibus—mortuis suis primis fundatoribus, viris sane apostolicis—ad propagandam fidem elegerunt sibi media pure humana illis directe contraria quibus Apostoli usi sunt, tamquam a Christo Domino nostro comendatis et post ipsos quotquot assumpti sunt ad regnum Dei evangelizandum, exceptis his patribus societatis.

16. [D. Navarrete], *The Travels and Controversies of Friar Domingo Navarrete 1618-1686*, edited by J.S. Cummins, 2 vols., (Cambridge: Cambridge University Press for the Hakluyt Society), 1:lxxvii; Cummins reports that Domingo overheard one Jesuit tell this to another.

Recent Catholic appraisal of the missionary efforts inspired by Ricci have been most laudatory. Pope John Paul II praised Ricci in an address commemorating the four hundredth anniversary of Ricci's arrival in China:

In speaking of the Gospel, he knew how to find the cultural mode that was appropriate to those who were listening to him. He began with a discussion of themes dear to the Chinese people, that is, morality and the rules of social life according to the Confucian tradition, the human and ethical values which he acknowledged with affection. Then he discreetly and indirectly introduced the Christian point of view on the various problems and matters. Thus, without imposing or commanding, he ended up by bringing many of his listeners to the explicit knowledge and to an authentic cult of God, the Supreme Good.

(Pope John Paul II, "Address on the Work of Father Ricci in China," *International Symposium on Chinese-Western Cultural Interchange in Commemoration of the 400th Anniversary of the Arrival of Matteo Ricci, S.J. in China*, [1983] Taipei, Taiwan, Republic of China, 3.)

17. H. Yi-long, "Court Divination and Christianity in the K'ang-hsi Era," *Chinese Science* 10 (1991): 1-20, has argued that issues regarding the proper methods for the practice of divination and hemerology (the determination of lucky days) were also included in the attacks on Schall and Verbiest.

18. A fuller study should be undertaken to enable a more complete understanding of the individuals and the context in which they worked. (Questions raised by the Reformation would have to be considered in a more complete treatment.)

I have purposely avoided discussion of yet another level of complexity: the whole question of the reaction to Galileo's work by the Jesuits in China. They were greatly interested in his work and attempted to correspond with him; Schall wrote the first account in Chinese of the telescope (1626). Certainly, a complete account of the Jesuit astronomers in China must deal with their reception of the work of Copernicus, Galileo, Tycho, and Kepler.

19. Brooke.

20. J.W. Draper, *History of the Conflict between Religion and Science* (London: n.p, 1875); and A.D. White, *A History of the Warfare of Science with Theology in Christendom* (New York: n.p., 1896); R.K. Merton, *Science, Technology and Society in Seventeenth-century England* (New York, Harper and Row, 1970; first published in *Osiris* 4 (1938), pt. 2, 360-632).

21. F.T. Chiang, "The History of Matteo Ricci's Missionary to China and the Meaning of His Book *The True Idea of God*," in *International Symposium on Chinese-Western Cultural Interchange in Commemoration of the 400th Anniversary of the Arrival of Matteo Ricci, S.J. in China* (Taipei, Taiwan, Republic of China, 1983), 138. M. Ricci, *The True Meaning of the Lord of Heaven (T'ien-chu Shih-i)*, translated by D. Lancashire and P.H. Kuo-chen, Chinese-English Edition edited by E.J. Malatesta, (St. Louis: The Institute of Jesuit Sources, 1985); see particularly the "Translators' Introduction," 33-35. Donald Wagner pointed out to me that *Shang-ti* could also be translated as "God on high."

22. Translation in Spence, *China Helpers*, 10: Original in A. Schall von Bell, *Lettres et Mémoires d'Adam Schall S.J. édités par le P. Henri Bernard S.J. Relation Historique*, edited and translated by P. Bornet (Tientsin: Hautes Études, 1942), 6-7:

Il n'y a rien d'étonnant s'ils ont aspiré de toutes leurs forces a réformer l'astronomie, étant donné que c'est une occupation purement scientifique et non déplacée pour des religieux. (6)

hoc priores Socii animadvertentes, nihil mirum si totis viribus ad restaurationem Astronomiae, cum res sit plane litteraria nec religiosis hominibus indigna, semper anhelaverint; (7)

pages 302-3 (translated in Spence, *China Helpers*, 22):

Je pensais en moi-même à tout le travail consacré depuis près de vingt ans à la réforme de l'astronomie et je voulais qu'il ne fut pas perdu, car c'est à lui que je devais et que j'espérais pour mes successeurs de pareils résultats. (En ce même temps, je voyais à ma grande douleur ce travail méprisé par ceux-là qui auraient dû surtout l'encourager.) (302)

viginti fere annorum Astronomiae restaurandae impensus labor quo mediante me vel posteros aliquando simile consecuturum sperabam, (quem laborem eo ipso tempore videbam potissimum conculcari ab iis, magno meo cum dolore, a quibus deberet promoveri).... (303)

23. [G.W. Leibniz], *The Preface to Leibniz' "Novissima Sinica,"* Commentary, translation, text by Donald F. Lach (Honolulu: University of Hawaii Press, 1957); Leibniz discussed the work of the Jesuits in chaps. 14-15. See also M.T. Ryan, 9-40, particularly 29-34.

24. Gernet, 193.

25. Gernet.

26. Gernet, 195-96.

27. H. Bosmans, "Ferdinand Verbiest, directeur de l'observatoire de Pékin (1623-1688)," *Revue des Questions Scientifiques* 71 (1912): 195-273 and 375-464, on 386. Translation in Spence, *China Helpers*, 33: "Une étoile conduisit jadis les rois mages à adorer le vrai Dieu; la science des astres conduira de même petit à petit les princes de l'Orient à connaître et à adorer leur Seigneur."

28. Translation in Cummins, 234. Original in F. Verbiest, *Correspondance de Ferdinand Verbiest de la Compagnie de Jésus (1623-1688), Directeur de l'Observatoire de Pékin*, edited by H. Josson and L. Willaert (Brussels: Palais des Académies [Commission Royale d'Histoire], 1938), 255: letter dated 1 September 1678: "Religio christiana per astronomiam in Sinas primum introducta, per astronomiam in Sinis conservata et cum a paucis annis fuit in exilium missa, per astronomiam revocata et in pristinam dignitatem restituta est; idea astronomiae brachiis merito incumbit."

3

Astronomy for the People: R.A. Proctor and the Popularization of the Victorian Universe[1]

Bernard Lightman

In 1882, Victorian popularizer of science Richard A. Proctor (1837–1888) published an entertaining introduction to astronomy titled *Easy Star Lessons*. In it Proctor covers the dominant constellations in the English night sky for each month, weaving into his discussion interesting tidbits of astronomical lore and stories of how constellations first received their names. Consonant with his aim of conveying to beginners the wonder of stargazing, Proctor emphasizes the role of imagination in the astronomer's craft. The ancients, he asserts, used their imagination to perceive a resemblance between a star-group and the figure after which they named it. Modern day astronomers are no less imaginative. "I have heard it said that the liveliest imagination cannot form figures of familiar objects out of the stars," Proctor reports.

> But this is certainly a mistake, for I know that when I was a lad, and before I learned to associate the stars with the constellations at present in use, I used to imagine among the stars the figures of such objects as I was most familiar with. In the constellation of the Swan, I saw a capital kite.[2]

The adult Proctor also invests the constellations with meanings "familiar," though more complex, to both his reading audience and himself. Just as evolutionists like Spencer imagined that they found in the organic world a basis for the status quo in Victorian society, Proctor, the popularizer of astronomy, found in the inorganic world scientific justification of a technological, middle-class social structure. Proctor's lively astronomical imagination projects onto the skies a universe composed of planets, inhabited by aliens organized into industrial societies blessed by God.

Until recently, scholars have paid little attention to Proctor and likely few readers will recognize the name. He was considered eminent enough by contemporaries for inclusion in *The Dictionary of National Biography*, however, and an entry on Proctor by J. D. North appears in *The Dictionary of Scientific Biography*.[3] In *The Extraterrestrial Life Debate*, Michael Crowe presents a persuasive case for the importance of Proctor, not only due to Proctor's participation in the pluralist debate, but also by virtue of the immense impact of approximately seventy books (the majority on astronomy), five hundred essays and eighty-three technical papers in the *Royal Astronomical Society Monthly Notices*. By the time of his death in 1888, Proctor had become, in Crowe's judgment, "the most widely read astronomical author in the English-speaking world."[4] Crowe reckons that "thousands" of members of the public were introduced to astronomy by Proctor's writings while North points out that a number of professional astronomers cut their teeth on one of Proctor's books.[5]

There are legitimate grounds for rescuing Proctor from obscurity in addition to his significance as popularizer. He contributed to the progress of late-nineteenth-century astronomy in his original scientific work on Venus and Mars. He charted the directions and proper motions of about 1,600 stars. And Proctor should be credited especially for his role in formulating the dominant cosmology of the period.

The youngest son of a wealthy solicitor, Proctor entered St. John's College, Cambridge in 1856 where he studied theology and mathematics. An early marriage to an Irish woman before his degree exam in 1860 was said to be the cause of his disappointing showing as twenty-third wrangler. Proctor then began to study for the bar while satisfying a growing interest in astronomy during his free time. Reading John Pringle Nichol's *Views of the Architecture of the Heavens* inspired him to write an essay entitled "Colours of the Double Stars," published in 1863 in *The Cornhill Magazine*. So great was his anguish over the tragic loss of his first child in 1863, that in order to distract himself from his suffering he took up the project of writing *Saturn and Its System* (1865) on the advice of his physician.

What began as a hobby and later served as a form of therapy became an economic necessity in 1866 when Proctor incurred a staggering debt of £13,000 as the result of the failure of a New Zealand bank in which he was the second largest shareholder. Though his literary career was never a resounding financial success, he was able to develop a writing style that eventually won him recognition from both professionals and the public. In 1866 he was elected to the Royal Astronomical Society and later filled the office of honorary secretary. Proctor's first major success, *Other Worlds Than Ours* (1870), was followed by triumphant lecture tours of America and Australasia.

After his first wife died in 1879, Proctor married an American widow two years later and settled at her hometown in St. Joseph, Missouri. In 1881 he founded the London scientific weekly *Knowledge and the Illustrated Scientific News*, which he edited and to which he contributed numerous essays, though he remained in the United States. He moved to Florida in 1887, but the following year, at the age of 51, he died of yellow fever in New York City while in transit to another lecture tour.

In the majority of his astronomical works, Proctor aimed at attracting the knowledgeable reading public, rather than the expert astronomer, and hoped to educate this audience as to the religious significance of the heavens. In the preface to his *New Star Atlas* (1872) he vows that "no pains have been spared to clear the maps of all which could cause confusion to the beginner."[6] *Easy Star Lessons* (1882) is intended to solve a problem which Proctor encountered as a boy when he tried to follow books of astronomy—the usual star charts gave no information about the places of star groups in the sky or the times during which they appeared. So Proctor's object in the book was "to remove this difficulty for young astronomers" by offering four maps per month for the entire year, each monthly set showing what stars could be seen towards the north, south, east and west at a certain convenient hour.[7] Proctor stressed hands-on astronomy, for he who takes his astronomy at second-hand from books:

> may lightly disregard the grand lesson which the heavens are always teaching, and find only the grotesque and the incongruous, where in reality there is the perfect handiwork of the Creator. But the astronomer, imbued with the sense of beauty and perfection which each fresh hour of world-study instils more deeply into his soul, reads a nobler lesson in the skies.[8]

Proctor therefore took great pride in his role of popularizer, for he was leading his readers to God through astronomy.

Crowe argues that Proctor was "the most widely read participant in the pluralist debate in Britain and America during the 1870-to-1890 period."[9] Throughout the earlier plurality of worlds controversy during the 1850s, the pluralist position was upheld by the evangelical Scottish physicist David Brewster (1781-1868), who could not countenance the idea of a God who would create a wasteful universe empty of life, except on earth. To what end, he asked, were there other worlds unless they were made to be the abode of life? Brewster's chief opponent was William Whewell (1794-1866), Master of Trinity, a Bridgewater Treatise author, and considered by his contemporaries to be an authority in numerous scientific fields. Though a liberal Anglican, Whewell refused to be open-minded about the possibility of extraterrestrial life. He fiercely defended the notion that earthly life was a special creation of a designing creator. The existence of wasted space presented no problems for natural theology in his mind.[10] Proctor entered the debate on the side of Brewster and those who believed that there was life on other planets. But he appreciated Whewell's position more and more as his pluralist views underwent a major transformation after 1870. As Proctor gave up on the inhabitability of planets in the solar system, one by one, thereby moving closer to Whewell, he adopted a theory of planetary evolution, which posited the notion that every planet at some time or another in its evolution, whether in the past, present, or future, is life-supporting.[11]

Sensitive to the charge that his theory of planetary evolution slighted God, Proctor maintained that the old fashioned view, which saw each orb in the solar system as the abode of life and each star as the center of a system filled with life, contained an "element of bitterness." The older view implied that before these planets became inhabited, there would have been universal lifelessness and that after they cease to be fit for habitation, there will be universal death. "But according to the view which I have suggested," Proctor announced,

> (the view, be it remembered, to which all the evidence points), though the number of habitable orbs be greatly diminished, it yet remains to all intent and purposes infinite, while, instead of regarding the duration of life in the universe as finite, we can perceive that it is eternal in time, even as it is infinite in spatial extension.[12]

Since planets throughout the universe are at different stages of development, life will always exist somewhere in the universe. Proctor sees his theory of planetary evolution as more in keeping with the conception of an omnipotent, omniscient and benevolent God who values life above all else. But in Proctor's mind his new theory was not only an improvement from a religious point of view; it also had the advantage of scientific superiority. It was the theory "to which all the evidence points."

Proctor believed that the rapid development of astronomical science since the time of Brewster and Whewell allowed for increasingly more informed decisions on controversial points. The study of extraterrestrial life was a legitimate concern of science and could be treated as scientifically as the analysis of strange lifeforms from the earth's past. Proctor was fond of pointing to the similarities between the astronomer's pluralism and the geologist's interest in fossils:

> Astronomy and Geology owe much of their charm to the fact that they suggest thoughts of other forms of life than those with which we are familiar. Geology teaches us of days when this earth was peopled with strange creatures such as now are found upon its surface. We turn our thoughts to the epochs when those monsters throve and multiplied, and picture to ourselves the appearance which our earth then presented.[13]

Though the astronomer cannot examine the actual substance of living creatures existing on other planets, or even their fossil remains, and though we have a dim conception of the conditions under which they live, "we see proofs on all sides that, besides the world on which we live, other worlds exist as well cared for and as nobly planned."[14]

Proctor maintained that the burden of proof should lie with those who denied the possibility of extraterrestrial life. "Until it has been demonstrated that no form of life can exist upon a planet," he declared, "the presumption must be that the planet is inhabited."[15] Proctor's certainty that the benefit of the doubt always went to pluralism was derived from his understanding of purpose in nature. For he believed that nature's "great end" is "to afford scope and room for new forms of life, or to supply the wants of those which already exist."[16] Wherever he turned, Proctor found new evidence of purpose in nature. When considering the glowing mass of Jupiter which can sustain no life, readers are invited to find a *raison d'être*, for Proctor cannot accept that God would create something for no purpose at all; the "wealth of design" in Saturn is so striking in Proctor's eyes that we cannot question but "that the great planet *is* designed for purposes of the noblest sort" though we may be unable to fathom those divine purposes. Proctor enthuses as if he were a Bridgewater Treatise author over the recent discoveries of science which "are well calculated to excite our admiration for the wonderful works of God in His universe."[17]

Proctor even structured *Other Worlds Than Ours* along the lines of a cosmic, post-Darwinian natural theology. The beginning chapters on "What Our Earth Teaches Us" and "What We Learn from the Sun" set the didactic tone for the entire book. Here nature's lessons concerning God's intentions and will are revealed by the telescope, spectroscope and the other tools of the astronomer's trade. These first two chapters are a part of a nine chapter section on the solar system, which leads into a series of three chapters on the stars and nebulae, extending the discussion of how God instructs us through nature to the rest of the universe. The concluding chapter, titled "Supervision and Control," deals with the lessons to be learned from an examination of astronomy and the providence of God.

Particularly revealing is Proctor's refusal to deal with what he calls the religious aspect of the plurality of worlds debate or to discuss his own personal views on the subject of revealed religion in this final chapter of the book. In other words, Proctor wanted to present a religious, yet nondenominational and uncontroversial, reading of God's relationship to the universe. According to Crowe, this strategy spared his *Other Worlds* from the attacks launched against some pluralist presentations.[18] However, it also conferred an ecumenical spirit upon his religious interpretation of astronomy attractive to Dissenters, Anglicans and even vaguely theistic scientific naturalists like Herbert Spencer, Edward Clodd and Grant Allen. Possibly, Proctor's nonsectarian position stems from his flirtation with Catholicism, to which he turned as he grieved for the loss of his son in 1863. Proctor apparently severed his connection in 1875 when he was told that some of his astronomical theories were not in conformity with the teachings of the Catholic Church.[19] Proctor likely decided to avoid presenting a sectarian version of his astronomical position, Catholic or otherwise, in order to reach the largest possible audience.

What Proctor does discuss in "Supervision and Control," the final chapter of *Other Worlds Than Ours*, concerns issues which fall "more naturally within the province of the student of science."[20] Included in this category are the concerns of the natural theologian: the omnipresence and infinite wisdom of God (divine supervision of the universe), and God's control of the universe through natural law, which raises the problems of miracles and human freedom. Proctor's aim here is to dispel any intellectual difficulties the reader may have as to God's strict supervision and control of his creation, and to clear the way for acceptance of the notion that every event in even the least important of God's worlds is "an essential part of the plan according to which all things were created from the beginning."[21] Proctor's deterministic, scientific analogue of the theological concept of Providence, justified through the most modern astronomical theories, is a fitting climax to a text designed to teach the public how to read the lessons to be found in the heavens above.

Proctor certainly looked upon his work in astronomy as strengthening the harmonious relationship between science and religion. He believed that theologians were responsible for disputes between science and theology, and stated that "*not a single dispute* between science and theology has *ever arisen* from scientific men proclaiming anything in regard to the bearing of new theories, or even of established truths, on belief in the supernatural."[22] Complaining of the literal-mindedness of theologians, Proctor charged that it was their insistence on a strict doctrine of inspiration which forced the devout Christian into the equally unacceptable alternatives of attributing scientific error to God in the first chapter of Genesis or swallowing the account as scientifically accurate. In comparing the Hebrew and Assyrian creation stories, both of which contain either an implicit or explicit system of science, Proctor notes that the Old Testament writer avoided statements "too definitely in accordance with scientific views which might in later ages be corrected" while the Assyrian account includes much which is inconsistent with modern science.[23] Contemporary theologians did not appreciate the wisdom of the biblical writer, who saw that his main role was to communicate moral and religious truths, not scientific ones.

In an article on "Science and Theology," Proctor launched another attack on Christian theologians, again insisting on a distinction between theology and authentic religion. Recalling his own training at Cambridge, where he passed through all the theological examinations required of those who are to become clergymen, Proctor denied that the training of theology brought the individual closer to God than the lessons of science did:

> Yet, so far as religious teaching is concerned, I can see nothing in anything I then learned that even approaches in its influence the effect of those studies that bring before the mind the infinite vastness of God's domain, its eternal duration, and the perfection of the laws which prevail throughout its whole extent alike in Space and Time. What is really held by many theologians to be anti-theological in science, is in reality that which makes the teachings of science most solemn and impressive.... Science, in fine, presents the Universe of God as aptly symbolising what we have been taught to consider the attributes of God Himself. It is this that so many theologians regard as anti-theological, because narrow theologies have pictured God after their own image, with which these infinite grandeurs are not consistent.[24]

In this passage Proctor vividly juxtaposes the devout conception of the universe offered by astronomical science to the narrow-minded anthropomorphisms of theology. Biblical literalism, combined with a tendency toward idolatry, led misguided theologians to impugn the discoveries of scientists and fan the flames of war between science and religion.

However, though Proctor's *Other Worlds* was intended as support for his views on the harmony between science and religion, the final chapter met with mixed reactions from reviewers. Indignant with Proctor's hesitation to express his views on controversial subjects, the anonymous reviewer in *The Quarterly Journal of Science* protested that

> when a writer says he will give us an insight into the nature and operations of the Almighty, but sees no advantage in making people uncomfortable by saying what he himself thinks on just those matters on which he is best able to form a judgment, his views of Divine action are not likely to be much heeded either by "believers" or "unbelievers."[25]

The English Mechanic rose to Proctor's defense and referred to his conclusion as a "bold, but reverent attempt to meet the religious difficulties" of pluralism, and predicted that most readers would "rise from its perusal a wiser and better man."[26] Though the reviewers disagreed on the success of the final chapter, the majority of them were receptive to Proctor's presentation of the extraterrestrial life issue within a religious framework. Recommending *Other Worlds Than Ours* to every lover of astronomy, *The Astronomical Register* declared that "unlike many scientific works, a tone of reverence towards the Creator of all things runs through the book, which is greatly to be commended."[27]

Leaving to one side the reaction of reviewers, to what can Proctor's success with the public as lecturer and writer be accounted? First, he possessed all the abilities required of a popularizer of science: skills at synthesizing details into a unified whole, tremendous powers of simplification, and a lucid and colorful style of presentation. Second, Proctor had an unerring instinct for highlighting those themes which captured the imagination of the audience. Proctor knew from the success of *Other Worlds Than Ours*, which became his most widely read book and continued in print until 1909, that the public was more intrigued with astronomical works if they contained a continuous discussion of the possible existence of extraterrestrial life. Close to the end of his life, Proctor looked back at his initial decision in the late 1860s to incorporate pluralist themes into all his publications. He candidly confessed that he took up the topic "not—let me admit now—for its own sake, but as a convenient subject with which to associate the results of scientific researches which I could in no other way bring before the notice of the general reading public."[28] Proctor's emphasis on the purposefulness in nature is connected to the third reason for his success. He rightly judged that casting his science into a teleological framework would appeal to an audience raised as children in Christian households.

However, Proctor was also popular because many of the lessons he drew from his teleologically interpreted astronomy were exactly what Victorian readers wanted to hear in the social, political and economic climate of the 1870s and 1880s. A case in point are the lessons dealing with the nature of society, not that Proctor was unique among scientists in turning to nature for instruction on how to structure society. Historians of Victorian science have discussed how Darwin and other middle-class scientists wishfully superimposed their ideal vision of society onto nature and then "scientifically" justified their notion of the "natural" quality of traditional social roles through an appeal to the natural order.[29] Like other middle-class scientists, Proctor conferred on nature an orderliness which he later used to justify a particular notion of social order. Proctor's teleological stories were particularly comforting to Victorians because they seemed to confer scientific and divine sanction to a fixed pattern for social life in a time of change and upheaval.

The sixties and seventies were the heyday of Social Darwinism. Both Proctor and Darwin were working in a turbulent social context which produced conditions congenial to the growth of a social philosophy which emphasized open competition, the wisdom of allowing the weak to die off, and the need for gradual, not revolutionary, change. During the last three decades of the nineteenth century the English economy stagnated as her industrial and economic global dominance was increasingly challenged by Germany and the United States. This period was also marked by urban and industrial unrest and the emergence of mass socialist working-class politics.[30]

The ideological dimension of Proctor's thought is intimately connected to his religiously inspired pluralism. Both depended on analogical argument for their scientific legitimacy. In his stimulating essay on experimental techniques in Victorian astronomy, Simon Schaffer has asserted that "the Victorian supposition that lab manipulations could generate knowledge of the Sun, stars and nebulae hinged on the security of the claim that terrestrial and celestial matters were relevantly similar."[31] Proctor was one of those astronomers who invested enormous time in shoring up the analogy between earthly processes and all celestial processes. In his chapter on Mars in *Other Worlds Than Ours*, Proctor insists that the existence of aqueous vapor in the Martian atmosphere leads to the conclusion that Mars has oceans and rivers—because the same natural processes which operate on earth also are at work on Mars. To Proctor, the lesson concerning Mars is clear and unassailable, as he has been guided "by no speculative fancies, but simply by sober reasoning."[32] Mars has oceans and rivers, which he maps in detail, and is hospitable to life.

But if it is safe to reason about the bodies in the solar system based on terrestrial analogy, is it valid to reason about the entire universe based on the analogy of the solar system? This is the question which Proctor confronts in chapter ten of *Other Worlds Than Ours*, titled "Other Suns Than Ours." Whereas the first nine chapters dealt with the solar system, chapter ten introduces a section of the book which discusses, in Proctor's own words, "the architecture of the universe."[33] Proctor counsels caution in light of the tremendous distances which separate the earth from even the nearest of the fixed stars. However he insists that we hold onto "the significant lesson taught us by the solar system" that "the teachings of analogy" are a sure and safe guide.[34] Appealing to the authority of Sir William Herschel, who believed that "in theorizing about the unknown, there can be no safer guide than the analogy of known facts," Proctor proceeds to claim universality for the limited empirical experience of earth-bound beings.[35]

Proctor was fond of drawing on new data generated by the spectroscope to support his position on analogy. To Proctor, the spectroscope provided an even more powerful tool than the telescope for overcoming the vast distances separating the astronomer from the objects of his study. Schaffer has pointed out that most astronomers engaged in spectroscopic research were also pluralists.[36] For Proctor, the spectroscope proved that the materials in the substance of the earth are similar to those in the sun, and that all the planets which circle round the sun are also constituted of the same materials. Furthermore, spectroscopic evidence pointed to a resemblance between our sun and all the stars. Since Proctor laid down as a general rule that planets in each system are constituted of the same materials which exist in the star round which they orbit, he concluded that all planets in the universe resemble the earth in their physical composition.[37] This train of thought led him to speculate on the nature of alien societies in the second chapter of the book on the sun:

> Now, we have in this general law [of resemblance between planets and their sun] a means of passing beyond the bounds of the solar system, and forming no indistinct conceptions as to the existence and character of worlds circling around other suns. For it will be seen, in the chapter on the stars, that these orbs, like our sun, contain in their substance many

of the so-called terrestrial elements, while it may not unsafely be asserted that all, or nearly all, those elements, and few or no elements unknown to us, exist in the substance of every single star that shines upon us from the celestial concave. Hence we conclude that around those suns also there circle orbs constituted like themselves, and therefore containing the elements with which we are familiar. And the mind is immediately led to speculate on the uses which those elements are intended to subserve. If iron, for example, is present in some noble orb circling around Sirius, we speculate not unreasonably respecting the existence on that orb—either now or in the past, or at some future time—of beings capable of applying that metal to the useful purposes which man makes it subserve. The imagination suggests immediately the existence of arts and sciences, trades and manufactures, on that distant world. We know how intimately the use of iron has been associated with the progress of human civilization, and though we must ever remain in ignorance of the actual condition of intelligent beings in other worlds, we are yet led, by the mere presence of an element which is so closely related to the wants of man, to believe, with a new confidence, that for such beings those worlds must in truth have been fashioned.[38]

With breathtaking confidence, Proctor here asserts the existence of extraterrestrial societies with cultures and economic structures similar to our Western industrial civilization due to the presence of iron and other familiar elements in all worlds throughout the universe. No matter how physically bizarre alien lifeforms may be, if they are rational beings, they will naturally build a society which resembles the English model.

Proctor returns to this theme later in the book in the chapter on "Other Suns Than Ours." Here, to the information supplied by the spectroscope respecting the materials of which the stars are composed, he adds evidence from the nature of stellar spectra which indicates that the stars supply heat to the worlds around them:

Thus the fact, that the stars send forth heat to the worlds which circle around them, suggests at once the thought that on those worlds there must exist vegetable and animal forms of life; that natural phenomena, such as we are familiar with as due to the solar heat, must be produced in those worlds by the heat of their central sun; and that works such as those which man undertakes on earth—works in which intelligent creatures use Nature's powers to master Nature to their purposes—must go on in the worlds which circle around Aldebaran, Betelegeux, around Vega, Capella, and the blazing Sirius.[39]

Here Proctor literally projects Western technological society onto the heavens. Bacon's injunction to use scientific knowledge to command nature, generally regarded as the origin of the Western technological impulse, must also be followed by intelligent beings on other planets—it is built into the very nature of things. Just as it is natural for stars to warm orbiting planets and make possible the existence of living things, and for the heat of stars to cause natural effects similar to those on earth, so it is natural and inevitable that alien societies develop science in order to manipulate nature for their own ends. These "works" or activities "must go on." What is (natural) on this earth (technology) must be in the universe. Evidence from the spectroscope, mixed with a teleological pluralism, became powerful resources for the presentation of technological society as the cosmic norm. The lesson as to what is the natural structure of any society built by living beings is clear. Socialists and others who object to industrialized, technological Western society pit themselves against the rest of the universe and God to boot.

For all Proctor's confidence on the testimony of astronomical science to the legitimacy of English middle-class society, his arguments rest on shaky grounds. Schaffer discussed the pressing need of experimental astronomers to establish a consensus that spectroscopes were untroubled instruments.[40] Effort had to be put into making the spectroscope perform its ideological work in proper fashion. Proctor had difficulty controlling his favorite astronomical tool. The ideological significance of information gathered from the spectroscope tended to vary with Proctor's own shifts in priorities. For example, in an essay "Other Habitable Worlds," published in 1868 in *St. Paul's Magazine*, and later republished in *The Orbs Around Us* (1872), Proctor draws a very different conclusion as to the resemblance of the sun and the stars. Here Proctor denies that stars resemble the sun in the proportions between elementary constituents. "Whereas the sun exhibits a close affinity to our own earth as respects the proportions which exist between its elementary constituents," Proctor writes, "the stars (centres, doubtless of other systems) exhibit no such affinity."[41] While Proctor stresses the difference between the sun and the stars in 1868, two years later in *Other Worlds* he emphasizes the similarities between the constitution of the sun and the stars in order to make his argument concerning the existence of extraterrestrial life organized in industrial societies. Proctor's priority in 1868 was to demonstrate the existence of life on planets within our solar system, hence his desire to stress the similarities between the sun and planets in the solar system and his willingness to concede a lack of resemblance between the sun and the stars. But his increasing allegiance to the theory of planetary evolution led him to give up life in each planet of the solar system and move toward accentuating the life scattered throughout the universe at different times and in different places. Evidence produced by the spectroscope could be subtly manipulated depending on how far one desired to extend the terrestrial analogy and the scope of pluralism.

The success of *Other Worlds Than Ours* rests in part on the incorporation of ideological and religious themes into Proctor's astronomical thought and not solely on the fascination of the public with the doctrine of pluralism. Proctor offered a divinely sanctioned, industrial cosmology which embedded technological capitalism within the very nature of things at a time when middle-class English society was perceived to be threatened by urban unrest, socialist radicalism, and foreign competition.

If Proctor's popularity stemmed from the way he reflected the concerns of his readers, a study of his work also reveals the continuing power of religion, even after the publication of Darwin's *Origin of Species*. John Brooke has pointed out that in the post-Darwinian period "natural theology had not yet lost its power to corroborate a preexisting faith."[42] It is particularly striking that Proctor's scientific theories and ideological position derived their authority from his religious thought. But the same could be said of Herbert Spencer's Social Darwinism, which was based upon the notion of a competitive society produced through the cosmic evolutionary process, which itself was guided by a mysterious unknowable deity.[43] Likewise, a new school of agnostics led by Charles Albert Watts, which arose in the 1880s, preached an evolutionary theodicy which legitimated a conservative vision of the social order.[44] In all three cases, cosmic scientific theories, authorizing a middle-class world, and driven by religious teleologies, appealed most to the Victorian public.

NOTES

1. The author would like to express his gratitude to Professor Michael Crowe and to Dr. Simon Schaffer for encouragement and helpful suggestions in investigating the astronomical thought of R.A. Proctor. Research for this project was undertaken with the support of a Social Sciences and Humanities Research Council of Canada Travel and Small Grants Award as well as funding from the Committee on Research, Grants and Scholarships, Faculty of Arts, York University.

2. R.A. Proctor, *Easy Star Lessons,* (New York: G.P. Putnam's Sons, 1888), 23.

3. E.M.C., "Proctor, Richard Anthony," in *Dictionary of National Biography*, edited by L. Stephen and S. Lee (London: Oxford University Press, 1959–1960), 16:419–21. J.D. North, "Proctor, Richard Anthony," in *Dictionary of Scientific Biography*, edited by C.C. Gillispie (New York: Charles Scribner's Sons, 1975), 11:162–63.

4. M.J. Crowe, "Richard Proctor and Nineteenth-Century Astronomy," *History of Science Society Meeting* 1 (1989): 1.

5. M.J. Crowe, *The Extraterrestrial Life Debate 1750–1900,* (Cambridge: Cambridge University Press, 1986), 377; North, 163.

6. R.A. Proctor, *A New Star Atlas,* (London: Longmans, Green and Company, 1872), vi.

7. Proctor, *Easy Star Lessons*, 9.

8. R.A. Proctor, *Other Worlds Than Ours* (New York: A.L. Fowle, 1870), 158.

9. Crowe, *The Extraterrestrial Life Debate*, 369.

10. J.H. Brooke, "Natural Theology and the Plurality of Worlds," *Annals of Science* 34 (1977): 221–86.

11. Crowe, *The Extraterrestrial Life Debate*, 373–74.

12. R.A. Proctor, "Life in Other Worlds," *Knowledge* 11 (1888): 231.

13. Proctor, *Other Worlds Than Ours*, 17.

14. Proctor, *Other Worlds Than Ours*, 18.

15. Proctor, *Other Worlds Than Ours*, 71.

16. Proctor, *Other Worlds Than Ours*, 32.

17. Proctor, *Other Worlds Than Ours*, 154, 159–60, 21.

18. Crowe, *The Extraterrestrial Life Debate*, 372.

19. Crowe, *The Extraterrestrial Life Debate*, 375.

20. Proctor, *Other Worlds Than Ours*, 307.

21. Proctor, *Other Worlds Than Ours*, 307.

22. R.A. Proctor, "The Book of Genesis," *Knowledge* 9 (1885): 29.

23. Proctor, "The Book of Genesis," 30.

24. R.A. Proctor, "Science and Theology," *Knowledge* 6 (1884): 475.

25. Anonymous, "Other Worlds than Ours." *Quarterly Journal of Science* 7 (1870): 372.

26. Anonymous, "Other Worlds than Ours," *English Mechanic* 11 (1870): 271, 241.

27. "Mr. Proctor's New Work," *Astronomical Register* 8 (1870): 144.

28. Proctor, "Life in Other Worlds," 231.

29. E. Richards, "Darwin and the Descent of Woman," in *The Wilder Domain of Evolutionary Thought*, edited by D. Oldroyd and I. Langham (Dordrecht: D. Reidel Publishing Company, 1983), 57; R. Young, *Darwin's Metaphor: Nature's Place in Victorian Culture* (Cambridge: Cambridge University Press, 1985).

30. Richards, 87-89.

31. S. Schaffer, "Where Experiments End: Table-Top Trials in Victorian Astronomy," in *Scientific Practise: Theories of Stories of Doing Physics*, edited by J. Buchwald (Chicago: University of Chicago Press, 1995), 275-299.

32. Proctor, *Other Worlds Than Ours*, 121.

33. Proctor, *Other Worlds Than Ours*, 231.

34. Proctor, *Other Worlds Than Ours*, 233.

35. Proctor, *Other Worlds Than Ours*, 234.

36. Schaffer, 275-299.

37. Proctor, *Other Worlds Than Ours*, 251.

38. Proctor, *Other Worlds Than Ours*, 59.

39. Proctor, *Other Worlds Than Ours*, 253.

40. Schaffer, 275-299.

41. R.A. Proctor, *The Orbs Around Us* (New York and Bombay: Longmans, Green and Company, 1902), 26.

42. J.H. Brooke, *Science and Religion* (Cambridge: Cambridge University Press, 1991), 317.

43. B. Lightman, *The Origins of Agnosticism* (Baltimore: Johns Hopkins University Press, 1987).

44. B. Lightman, "Ideology, Evolution and Late-Victorian Agnostic Popularizers," in *History, Humanity and Evolution*, edited by J. Moore (Cambridge: Cambridge University Press, 1989), 285–309.

4

The Role of Belief in Modern Cosmology[1]

John Byl

INTRODUCTION

Cosmology is concerned with describing and explaining the universe as a whole. The object of this paper is to examine the limits of cosmological knowledge. In particular, I shall be concerned with discerning the role of belief in modern cosmology. How much of cosmology is based on belief? How do religious beliefs influence our cosmology? From a Christian point of view, how much of a role should be played by Christian beliefs? What, if anything, does the Bible have to contribute to our understanding of the cosmos? These are some of the questions we shall address.

Cosmological knowledge is more difficult to acquire than that of most other sciences. Part of the problem arises because there is only one universe to observe; hence we cannot compare it with similar objects and infer its probable nature. Moreover, we can observe it from only one small region in space-time. Even then, our access to celestial objects is relatively indirect, being limited to radiation emitted (presumably) from them. In explaining our observations, we are further restricted in that we can apply only knowledge acquired under the limited conditions of local laboratories.

Thus we are faced with the problem of extending local physics to limited astronomical observations in order to derive a model of the universe. This cannot be done without first making some theoretical assumptions about the nature of the universe. Some of these may be inherently unverifiable. Since the astronomical observations and local physics can be extended in various ways, our cosmology is dependent upon the assumptions made. Therefore the fundamental problem is that of choosing and justifying an appropriate set of presuppositions needed to construct a model of the universe. In this, we are strongly dependent upon our prior philosophical and religious prejudices as to how the universe should behave.

SOME BASIC ASSUMPTIONS

Assumptions in cosmology include such broad ones as the assumption of the universal validity of local physics (particularly general relativity), the assumption that we occupy a typical position in the universe, and the assumption that the universe can be represented by a four-dimensional space-time continuum.[2] Further, more detailed suppositions are also made. For example, the galactic redshifts are assumed to be due to the expansion of the universe, and the present universe is assumed to have evolved out of a past singularity.

Induction

It is generally assumed that the principle of induction is valid: that the laws of physics observed here and now are universally applicable, and that explanations of structure are to be given in terms of these laws. While such uniformity principles may seem reasonable enough, they are not unproblematic. The justification of induction has been one of the outstanding problems in the philosophy of science and is now widely considered to be insoluble. As David Hume pointed out in 1739, there is no compelling reason for believing that the principle of induction is valid. Induction cannot be justified by observation (since the unobserved universe is, by definition, unobserved) nor by logic (since there is no logical reason why the universe must behave uniformly); hence the universe beyond our experience may be quite different from what we might expect.

Induction is a particular problem for cosmology, since it strives to depict the entire history of the entire physical universe. The problem is not merely one of assuming that the laws discerned here and now are universally applicable. It is further assumed that laws valid under quite limited local conditions will still apply under vastly different circumstances such as near the Big Bang singularity.

As an extreme example of an alternative model, consider the drastic notion that the universe was recently created instantaneously. In the words of cosmologist George Ellis:

> A modern cosmologist who is also a theologian with strict fundamentalist views could construct a universe model which began 6000 years ago in time and whose edge was at a distance of 6000 light years.... A benevolent God could easily arrange the creation...so that suitable radiation was travelling toward us from the edge of the universe to give the illusion of a vastly older and larger expanding universe. It would be impossible for any other scientist on the earth to refute this world picture experimentally or observationally; all that he could do would be to disagree with the author's cosmological premises.[3]

Such an apparently radical view has a number of positive attributes. Since it refers to the past, no present or future observations or experiments can refute it. Nor is there anything illogical about such an origin of the universe. Another physicist, Herbert Dingle, writes of this theory:

> There is no question that the theory is free from self-contradiction and is consistent with all the facts of experience we have to explain; it certainly does not multiply hypotheses beyond necessity since it invokes only one; and it is evidently beyond future refutation. If, then, we are to ask of our concepts nothing more than that they shall correlate our present experience economically, we must accept it in preference to any other. Nevertheless, it is doubtful if a single person does so.[4]

One might object that such theories are untestable, and hence not scientific. However, physicist Frank Tipler has shown that it is possible to construct falsifiable creationist models:

> It is universally thought that it is impossible to construct a falsifiable theory which is consistent with the thousands of observations indicating an age of billions of years, but which holds that the Universe is only a few thousand years old. I consider such a view to be a slur on the ingenuity of theoretical physicists: we can construct a falsifiable theory with any characteristics you care to name.[5]

The basic thrust of his model is that, while the universe may appear to be very old, this is just an illusion.[6] Such an illusionary history is not unique to his theory. The many-worlds interpretation of quantum mechanics requires that, due to the observed interference of probability amplitudes, there are in reality many alternative histories that give rise to the present. This view requires that the existence of historical records does not imply that any past event has occurred. Although

Tipler claims not to believe his theory, he states that he developed it to challenge cosmologists and philosophers to give good reasons for rejecting it on scientific grounds. He asserts that his theory satisfies not only falsifiability, but most other criteria discussed in the scientific literature.

From a Christian point of view, if God is omnipotent, it is at least possible that he could have created the universe instantaneously. While such a theory may be difficult to disprove on scientific grounds, the prime objection to such models is the theological one that it seems to involve deception by God. However, proponents for this view counter this objection by asserting that since God himself has revealed a recent *creatio ex nihilo* he can hardly be accused of deception.[7] Clearly, one's beliefs about God and nature contribute significantly to the selection of one's cosmology.

The Cosmological Principle

A second assumption commonly made concerns an observational feature: the universe about us appears to be remarkably "isotropic," that is, it looks roughly the same in all directions. One obvious explanation is that we are near the center of a spherically symmetric universe. But such a solution is repugnant to modern cosmologists. As Ellis remarks:

> In ages by, the assumption that the Earth was at the centre of the universe was taken for granted. As we know, the pendulum has now swung to the opposite extreme; this is a concept that is anathema to almost all thinking men....It is due to the Copernican-Darwinian revolution in our understanding of the nature of man and his position in the universe. He has been dethroned from the exalted position he was once considered to hold.[8]

It would certainly be consistent with present observations that we are at the center of the universe, and that, for example, radio sources are distributed spherically and symmetrically about us in shells characterized by increasing source density and brightness as their distance from us increases. Although mathematical models for such Earth-centered cosmologies have occasionally been investigated, they have not been taken seriously; in fact, the most striking feature of published discussions of radio sources is how this obvious possibility has been completely discounted.

Instead, to explain the observed isotropy, the Cosmological Principle is adopted. This presupposes that we occupy a typical, rather than a special, position in the universe; it is assumed that all hypothetical observers throughout the universe would, at the same cosmic time, observe the same isotropic features of the universe. This implies that the universe can have no edges: either the universe is a finite, spherically-curved space, or it is infinite. Since we can observe the universe from only one position—our own—there can be no direct evidence for the Cosmological Principle. Yet there is an indirect test: if the Cosmological Principle holds, then the universe should be spatially homogeneous: at any given time the distribution of matter should be roughly the same throughout the universe.

The observations, however, indicate that the distant galaxies are not distributed uniformly in space. To some degree this may be expected, since the more distant galaxies presumably represent an earlier epoch when the universe was denser and the galaxies were closer together. Even after correcting for this effect, however, the density of galaxies appears to be a function of their distance from us. At first sight this would seem to refute the Cosmological Principle. Nonetheless, it is saved from falsification by postulating that galaxies evolve in time; in the past the galaxies were not merely closer together, but more galaxies were in existence than there are now. The evolution rate is not determined independently, but is adjusted so as to make the universe homogeneous:

> The assumption of spatial homogeneity has inevitably been made, and has led to the conclusion that the population of radio sources evolves extremely rapidly. What has therefore happened is that an unproven cosmological assumption has been completely accepted and has been used to obtain rather unexpected information about astrophysical processes.[9]

In short, the Cosmological Principle is a metaphysical belief which is saved from falsification by the introduction of ad hoc auxiliary theories.

The Cosmological Principle does have the advantage of yielding a relatively simple model. Convenience alone is not sufficient to demonstrate veracity, however. It is possible to construct other models based upon different assumptions. For example, steady-state cosmology is based on the Perfect Cosmological Principle: the assumption that the universe is roughly the same not only in space *but also in time.* One could also drop the Cosmological Principle altogether and build models that place us near the center of a spherically symmetric universe.[10]

Similarly, one could question other assumptions such as, for example, the Big Bang origin of the microwave background radiation and the notion that the galactic redshifts are caused by the expansion of the universe. It should be noted that the redshifts do exhibit some observational features that argue against the expansion explanation: the redshifts of nearby galaxies seem to be bunched together in regular intervals,[11] and Arp[12] has documented many cases where galaxies that seem to be physically connected have widely differing redshifts. In recent times a host of alternative cosmologies have been presented,[13] some even positing a static universe.[14]

THE PROBLEM OF VERIFICATION

Some of the most basic assumptions in cosmology are of an essentially unverifiable nature. Verification can also be a problem for specific aspects of cosmological models. Oldershaw distinguishes between two types of untestability: (1) a theory that cannot generate definitive testable predictions or whose predictions are impossible to test is *inherently* untestable ("untestability of the first kind"); (2) a theory that has many adjustable parameters is *effectively* untestable ("untestability of the second kind").[15]

Many of the basic features of Big Bang cosmology, the currently favored model, are inherently untestable. The most critical events supposedly occurred within 10^{-25} seconds after the Big Bang. Yet, in principle, we cannot obtain direct information on the state of the universe prior to the decoupling of radiation and matter at 10^{-13} seconds after the Big Bang.[16] The latest inflationary Big Bang models are heavily dependent upon particle physics, which in turn involves more unverifiable theoretical entities. Many theories of the new physics require extra dimensions: 5 to 26 dimensions is typical and about 950 dimensions is the latest record. Yet there is no known way to test empirically for the existence of these extra dimensions.[17] Further, the conditions in the early universe (tremendously high temperatures and pressures) cannot be reproduced elsewhere, also making particle physics inherently untestable.

There are numerous cases involving untestability of the second kind. Particle physics has been applied to overcome various observational shortcomings of Big Bang cosmology. Yet the standard model of particle physics has more than 20 parameters (such as particle masses and coupling strengths of the forces) that cannot be uniquely derived and are thus freely adjustable. There are currently at least half a dozen superstring theories. Many of the problems in particle physics are "solved" ad hoc by inventing new concepts, such as the "Higgs mechanism," renormalization, "color," etc.[18] P.J.E. Peebles has wryly remarked:

> The big news so far is that particle physicists seem to be able to provide initial conditions for cosmology that meet what astronomers generally

think they want without undue forcing of the particle physicist's theory. Indeed I sometimes have the feeling of taking part in a vaudeville skit: "you want a tuck in the waist? We'll take a tuck." You want a massive weakly interacting particle? We have a full rack...This is a lot of activity to be fed by the thin gruel of theory and negative observational results, with no prediction and experimental verification of the sort that, according to the usual rules of physics, would lead us to think that we are on the right track....[19]

To explain the recently-discovered large-scale structure in the universe there are at least three theories: superconducting cosmic strings, biased galaxy formation in a WIMP-dominated universe, and double inflation.[20] Numerous theories have also been proposed to explain the supposed ninety percent of the "missing mass" of the universe.

CRITERIA

It is clear that the observed celestial phenomena can be explained within the frameworks of a variety of different models. How are we to choose?

Within the last half century it has come to be generally accepted that the origin of scientific theories is largely subjective. For example, Karl Popper concluded that "We must regard all laws or theories as hypothetical or conjectural; that is, as guesses"[21]; he saw theories as "the free creations of our minds."[22] As Carl Hempel puts it: "The transition from data to theory requires creative imagination. Scientific hypotheses and theories are not derived from observed facts, but are invented in order to account for them."[23] It seems that theories are not so much *given to* us by nature as *imposed by* us on nature; they are not so much the result of rational thought as they are the creations of our intuition. Of course, there is some constraint on our theorizing, since our theories must be consistent with observation, but this still leaves us with many possibilities.

While one might think that further research may falsify the majority of such theories, this is not easily done. A favored theory, such as Big Bang cosmology, can always be saved from observational falsification by adding suitable supplementary hypotheses (for example, inflation, Higgs mechanisms, multi-dimensions, various esoteric forms of "dark matter," etc.). A theory that must be supported by artificial, ad hoc devices is not plausible. Nevertheless, however difficult it may be to demonstrate a particular ad hoc theory to be true, it is even harder to conclusively disprove it. According to Imre Lakatos: "Scientific theories are not only equally unprovable, and equally improbable, but they are also equally undisprovable."[24]

Popper objected to the notion that a favored theory could be continually adjusted to avoid observational disproof. Recognizing that there was no logic to the discovery of theories, he hoped to construct a rational methodology for the objective selection of theories. He proposed that genuine scientific theories should be falsifiable: they should make definite testable predictions which, if unfulfilled, would cause the theory to be discarded. However, if we were to apply this criterion to cosmology, we would have little theory left over: virtually all current cosmological models have been falsified by observations. Nor does Popper offer any justification as to why easily falsifiable theories are more likely to be true than others.

It is, of course, possible to play the game of cosmology under different rules. Lakatos described science as consisting of competing research programs, rather than consisting of competing theories. In this view, ad hoc modifications are permitted as long as they are in accord with the basic theoretical principles that are guiding the research program. Various criteria for assessing theories—or research programs—have been suggested. For example, astronomer Howard van Till lists those of cognitive relevance, predictive accuracy, coherence, explanatory scope, unifying power, and fertility.[25] Yet, while such criteria may seem plausible enough, it is generally acknowledged that they are by no means rigorous and merely reflect values used in practice. Indeed, the creation of selection criteria is no less subjective than the creation of scientific theories. As Lakatos notes:

> These scientific games are without genuine epistemological content unless we superimpose on them some sort of metaphysical principle which will say that the game, as specified by the methodology, gives us the best chance of approaching the truth.[26]

Science in general is plagued by the lack of definite, objective criteria that might enable us to readily distinguish true theories from false ones. While theories should, of course, be consistent with observational facts, these still underdetermine scientific theories. It is at this crucial point—choosing preferred theories—that we must generally rely on extrascientific factors.

THE ROLE OF RELIGIOUS BELIEFS

This brings us to the question as to what role religious commitments should play in cosmological theorizing. Religious beliefs have affected the creation, assessment and selection of cosmological theories. For example, Fred Hoyle rejects Big Bang cosmology at least in part because of its perceived theistic implications,[27] while Christians such as William Craig and Hugh Ross use the Big Bang as evidence for God.[28] Furthermore, the rejection by creationists of a long evolutionary history of the universe is based primarily on religious commitments. Finally, the National Academy of Science in the U.S.A. has objected to creationism mainly on the grounds that creationism "subordinates evidence to statements made on authority and revelation" and that "it accounts for the origin of life by supernatural means."[29] Yet modern cosmology is marked by a pervasive naturalism that leaves little room for any religion in the traditional sense. However much one might wish to eliminate the supernatural from science, it is quite another matter to prove that it actually is absent from reality.

Van Till has contended that extrascientific dogma should not influence our assessments and selection of theories:

> Religious commitments, whether theistic or nontheistic, should not be permitted to interfere with the normal functioning of the epistemic value system developed and employed within the scientific community. Great mischief is done when extrascientific dogma is allowed to take precedence over epistemic values such as cognitive relevance, predictive accuracy....Science held hostage to any belief system, whether naturalistic or theistic, can no longer function effectively to gain knowledge of the physical universe....Science held hostage by extrascientific dogma is science made barren.[30]

His exclusion of religious commitments from the selection of scientific theories follows from the complementarity of science and religion. But even this view is itself based on extrascientific considerations. At heart we cannot avoid being guided by religious and philosophical factors in our assessment and selection of theories. Religious and philosophical prejudices, however, have at times blinded their adherents to blatant deficiencies in favored theories and to obvious advantages in rival models. Therefore, it is important to minimize undue distortion and bias by stating our premises and criteria openly.

For Christians the question is how to rate divine revelation. If an omniscient God has revealed truth to us, it seems reasonable that divine revelation should speak authoritatively on all it addresses. Such a divine source of knowledge would carry more epistemological weight than speculative theorizing, with its subjectivity and fallibility. For example, some hold that the Bible is the inerrant word of God and that the traditional interpretation of the Bible is the intended one. Scripture, thus defined, and observation could be given an equal rank, both being assumed to be of divine origin, with scientific theorizing given a much lower rating. Cosmological theories could be selected in conformity to both Scripture and observation. Such a fundamentalist approach to cosmology is no less scientific than other approaches; it is merely based on different philosophical presuppositions. This could well result in the rejection of most modern cosmogonies in favor of, say, the instantaneous universe discussed above. Alternatively, one could refuse to speculate about matters beyond Scripture and observation, adopting, for example, an instrumentalist approach to scientific theorizing.

Ernan McMullin has objected to such a strong form of what he terms the "relevance-of-theology-to-cosmology principle."[31] He asserts that Scripture has no direct cosmological intent. To this one could respond that, while the Bible admittedly has little to say about the physical structure of the universe, it certainly does address such relevant issues as the origin and destiny of the universe, as well as pointing towards the existence of a spiritual realm.

RELIGION AND MODERN COSMOLOGY

The impact of religion on cosmology, however, goes beyond the selection of theories. Some of the issues at stake are greater than those involving origins and concern the nature of God and the question of life after death.

God and the Universe

A most basic question concerns God's relation to the physical universe. In recent years a number of cosmologists have depicted God as a natural, evolving being who is very much part of the physical universe.[32] Davies pictures God as a supermind who can "load the quantum dice," thereby controlling everything that happens, and thus escaping our attention. Neither P. Davies nor F.J. Dyson explain how this supermind, who really does not come into his own until the far future, could have influenced the initial conditions and subsequent evolution up to now.

This deficiency has been addressed with the idea that evolution is controlled by an evolved intelligence placed infinitely far in the future, its present influence dependent upon its subsequent future status through what Hoyle calls "backward causation."[33]

F.J. Tipler defines life as information processing. As the Big Crunch is approached, life will become omnipresent, omnipotent (that is, it will control all matter and energy sources), and omniscient (that is, the information stored will be infinite). To ensure that we do arrive at this final singularity—the "Omega Point," Tipler proposes that the wave function of the universe is such that all classical paths terminate in a future Omega Point, with life coming into existence along at least one classical path and continuing on to the Omega Point. In a sense the Omega Point creates the physical universe, but in another sense the Omega Point creates itself. Everything is predetermined by the wave function of the universe. However, the origin of the wave function itself is not explained. This scheme resembles Hoyle's backward causation.

Whereas the above authors believe their conception of a natural god is more plausible than the traditional God of Christianity,[34] it is doubtful whether such gods would satisfy many believers. These natural gods, subject as they are to natural law (except for backward causation), can perform no miracles, answer no prayers, and have few of the characteristics generally attributed to the God of the Bible. Moreover, since they could not have existed prior to the (presumed) Big Bang singularity and will not evolve into superminds for a long time, their past and present influence can be brought about only through backward causation, which seems to be a form of supernaturalism in disguise.

However, a number of attempts have been made to construct gods more in accord with Christianity. For instance, Teilhard de Chardin viewed Christ as the Omega Point, the ultimate goal of the evolutionary process.[35] Process theology, as exemplified in the work of Alfred North Whitehead and his followers, also considered God as an evolving being.[36] The world is seen as the body of God, but this God also has a mind which is, however, dependent upon his body. Creatures in the universe are considered as cells of God's body. Both God and the universe are eternal, God having created the world out of preexisting material. God is generally thought to be omniscient with respect to all past and present events, but not with regard to the future. The future is indeterminate and not even God can know it.[37] Nor is this God omnipotent. God's acting does not contradict science; God always acts with and through other entities rather than as a substitute for their actions by acting alone. Process theologians generally reject miracles.

While such concepts of God are often presented within a Christian tradition, they fall short of the God of orthodox Christianity, who is omnipotent, omniscient, transcendent (he stands above and is independent of his creation), and is spirit (he is not limited to physical reality). This God can be incorporated into our cosmology by considering the universe as larger than our three-dimensional physical cosmos. Luco van den Brom suggests that God exists spatially in his own higher-dimensional universe, and created a three-dimensional world in his more-dimensional world.[38] Van den Brom views the ascension as the withdrawal of Christ's body from the three-dimensional created world into the higher-dimensional system of heaven. This higher-dimensional reality is somewhat similar to Karl Barth's "superspace" and Karl Heim's "suprapolar space," except that van den Brom attaches more concrete reality to his space than the other two seem to attach to their spaces. John Polkinghorne also postulates the existence of another realm—the "noetic" world—which includes not only mental entities, such as mathematical truths, but also spiritual entities such as angels.[39]

Van den Brom's higher-dimensional space makes it possible to speak of God's heaven as a place outside of our space without having to consider heaven as a place in an unreal sense. We could consider heaven as having more than three dimensions, or as being a three-dimensional world parallel to our own in a four-dimensional space, much like two two-dimensional planes embedded in a three-dimensional space. In such a space it is also natural that our physical world could be influenced by factors outside of it. The higher-dimensional world of God could have its own laws alongside of the laws of our three-dimensional world. These higher laws and dimensions are not open to scientific research. Miracles could be explained as the intrusion of higher-dimensional factors into our three-dimensional world.

The Future: Life After Death

Beliefs in cosmology concern not only the past and the extent of the universe. A crucial theological question concerns life after death. The central hope of Christianity is that of the return of Christ, the last judgment, and eternal reward or punishment. The Bible speaks of the creation of a new heaven and a new earth. And these events will occur relatively soon—not billions of years in the future.[40] Thus biblical eschatology is rather different from that of modern cosmology, which offers little hope for the future: neither for individual immortality, nor even for the survival of life in any form.

None of the natural gods discussed above can grant us an individual, conscious life hereafter (that is, subjective immortality). Life may survive collectively, but this is of little consolation to the individual. Most process theologians reject the notion of any individual immortality. According to Ogden the human person will continue to live on only in God's cosmic memory, of which we will not be conscious.[41] A similar position is taken by Charles Hartshorne, who considers the notion of an actual heaven and hell to be a dangerous error.[42] According to process theologian John Cobb, a major difficulty in the belief in the separation of body and soul is where to place the soul; we can no longer conceive of heaven and hell as spatial places.[43]

In Newtonian cosmology, souls or mental substances fit in so ill with the space-time continuum that it did not seem too strange to postulate another sphere, a spiritual realm, where human souls belonged. But in the evolutionary cosmos this distinction between mind and matter cannot be maintained. If minds emerged in the physical universe then they must belong in that universe. There seems to be no longer a "place" for life after death.

Even Teilhard affirmed that at death our body decomposes and our soul, being tied to the body, cannot survive as a high order of consciousness. Although Teilhard brings Christ into the picture, it is a much diminished Christ who offers no eternal salvation. The Dutch astronomer Herman Zanstra concludes that Teilhard's rejection of the immortality of the soul does not allow for a true religion in the full sense. According to Zanstra, the chief problem for religion is whether the soul exists independent of the body without perishing.[44] If not, there would be no room for God as a Spirit; such parallelism means atheism. Forms of complementarity or consonance between naturalistic cosmologies and Christianity break down because they exclude spiritual reality. Anyone who believes in a conscious and individual subjective existence after death must reject naturalistic cosmologies as an adequate description or explanation of reality. Zanstra, therefore, opts for dualism with interaction: soul and body are separate entities which influence each other, yet have a degree of independence. His supernatural view of the world includes conscious spirits, where consciousness has existed before the physical universe began, and will continue to exist while our universe is reduced to dust and ashes.

It seems, then, that an essential ingredient for a religion which posits both a supernatural God and subjective immortality is the existence of a spiritual realm wherein God and soul can exist. Van den Brom suggests that the physical universe is a subspace of a larger more-dimensional universe. If the latter is spiritual rather than physical, the contents of the Bible would fit very readily.

CONCLUSIONS

In summary, adequate, objective criteria for selecting theories are lacking in modern cosmology. Cosmological theories are speculative, ad hoc and not easily testable. Therefore, a large role is played by philosophical and religious beliefs, and Christians must not permit modern cosmology to unduly modify their religious beliefs. On the contrary, they should hold on to the faith, construct a cosmology consistent with it, and look forward with confidence to the return of Christ.

In such a Christian approach to cosmology, conformity to Scripture should be a prominent criterion for theory selection. For instance, belief in a transcendent God carries with it a belief in a reality larger than our physical space-time. Allowance for a spiritual dimension to reality opens up the possibility of nonphysical causes for physical events. God is free to act in miraculous ways. This in turn leaves open the possibility of instantaneous creation, miracles, the Incarnation, Resurrection, and life after death.

NOTES

1. I wish to thank the Social Sciences and Humanities Research Council of Canada (Aid to Small Universities Program, Grant No. 481-90-0032) for funding in support of the research and presentation of this paper.

2. See, for example, the discussion by W. Stoeger, "Contemporary Cosmology and Implications for the Science-Religion Dialogue," in *Physics, Philosophy and Theology: A Common Quest for Understanding*, edited by J.R. Russell, W.R. Stoeger and G.V. Coyne (Vatican City: Vatican Observatory Press and Notre Dame: University of Notre Dame Press, 1988), 219-47.

3. G.F.R. Ellis, "Cosmology and Verifiability." *Quarterly Journal of the Royal Astronomical Society* 16 (1975): 246.

4. H. Dingle, "Philosophical Aspects of Cosmology," *Vistas in Astronomy* 1 (1960): 166.

5. F.J. Tipler, "How to Construct a Falsifiable Theory in which the Universe Came into Being Several Thousand Years Ago," in *Proceedings of the 1984 Biennial Meeting of the Philosophy of Science Association: Volume 2*, edited by P. Asquith and R. Giere (East Lansing: Philosophy of Science Association, 1984), 873.

6. Tipler's theory, involving retrodiction barriers caused by exploding black holes, is rather technical. For further details the interested reader is referred to his paper.

7. E.H. Andrews, "The Age of the Earth," in *Creation and Evolution: When Christians Disagree*, edited by O.R. Barclay (Downers Grove: InterVarsity Press, 1985), 64.

8. Ellis, "Cosmology and Verifiability," 250.

9. Ellis, "Cosmology and Verifiability," 250.

10. See, for example, the two-centred model of G.F.R. Ellis, "Is the Universe Expanding?" *General Relativity and Gravitation* 9 (1978): 87-94. A

more recent model is that of J.K. Rao and M. Annapurna, "Spherically Symmetric Static Inhomogeneous Cosmological Models," *Pramana* 36 (1991): 95-103.

11. For a popular account see J. Gribbon, "'Bunched' Redshifts Question Cosmology," *New Scientist* 132 (1991): 1800-1.

12. H.C. Arp, *Quasars, Redshifts and Controversies* (Berkeley: Interstellar Media, 1987).

13. See, for example, G.F.R. Ellis, "Alternatives to the Big Bang," *Annual Review of Astronomy and Astrophysics* 22 (1984): 157-84.

14. For example, Ellis "Is the Universe Expanding?"; P.A. La Violette, "Is the Universe Really Expanding?" *Astrophysical Journal* 301 (1984): 544-53; P. Marmet and G. Reber, "Cosmic Matter and the Nonexpanding Universe," *Institute of Electrical and Electronics Engineers (IEEE) Transactions on Plasma Science* 17 (1989): 264-69.

15. R.L. Oldershaw, "The New Physics—Physical or Mathematical Science?" *American Journal of Physics* 56 (1988): 1075-81.

16. Oldershaw, 1077.

17. Oldershaw, 1077.

18. Oldershaw, 1078.

19. P.J.E. Peebles, "Physics of the Early Universe," *Science* 235 (1987): 372.

20. WIMP stands for a hypothetical class of weakly interacting particles.

21. K.R. Popper, *Objective Knowledge: An Evolutionary Approach* (London: Oxford University Press, 1972), 9.

22. K.R. Popper, *Conjectures and Refutations* (London: Routledge and Kegan Paul, 1963), 192.

23. C.G. Hempel, *Philosophy of Natural Science* (Engelwood Cliffs: Prentice-Hall, 1966), 15.

24. I. Lakatos, *The Methodology of Research Programmes* (Cambridge: Cambridge University Press, 1980), 19.

25. H.J. Van Till, "The Character of Contemporary Science," in *Portraits of Creation*, edited by H.J. Van Till, R.E. Snow, H.H. Stek and D.A. Young (Grand Rapids: William B. Eerdmans Publishing Company, 1990), 141-46.

26. Lakatos, 122.

27. F. Hoyle, *Astronomy and Cosmology* (San Francisco: Freeman, 1975), 684-85.

28. W.L. Craig, "Philosophical and Scientific Pointers to Creation *ex nihilo*," *Journal of the American Scientific Affiliation* 32 (1980): 5-13; H. Ross, *The Fingerprint of God* (Orange: Promise Publishing Company, 1991).

29. *Scientific Creationism: A View from the National Academy of Science*, (Washington: National Academy Press, 1980).

30. Van Till, 149-50.

31. E. McMullin, "How Should Cosmology Relate to Theology?" in *The Sciences and Theology in the Twentieth Century*, edited by A.R. Peacocke (Stocksfield: Oriel Press, 1981), 17-57.

32. P. Davies, *God and the New Physics* (New York: Simon and Schuster, 1983); F.J. Dyson, *Infinite in All Directions* (New York: Harper and Row, 1988).

33. F. Hoyle, *The Intelligent Universe* (New York: Holt, Rinehart and Winston, 1984); F.J. Tipler, "The Omega Point Theory: A Model of an Evolving God," in *Physics, Philosophy and Theology: A Common Quest for Understanding*, edited by J.R. Russell, W.R. Stoeger and G.V. Coyne (Vatican City: Vatican Observatory Press and Notre Dame: University of Notre Dame Press, 1988), 313-31; F.J. Tipler, "The Omega Point as Eschaton," *Zygon* 24 (1989): 217-53.

34. Dyson, 211.

35. P.T. de Chardin, *The Phenomenon of Man* (London: Collins, 1959).

36. A.N. Whitehead, *Process and Reality* (New York: Macmillan Publishing Company, 1992).

37. R.N. Nash, *The Concept of God* (Grand Rapids: Academe Books, 1983), 27-28. This book contrasts process theology with classical theism.

38. L.J. van den Brom, *God Alomtegenwoordig* (Kampen: J.H. Kok, 1982).

39. J. Polkinghorne, *Science and Creation* (London: Society for Promoting Christian Knowledge, 1988), 76.

40. Tipler, "The Omega Point Theory," 316.

41. S. Ogden, "The Meaning of Christian Hope," *Union Seminary Quarterly Review* 30 (1975): 160-63.

42. C. Hartshorne, *The Logic of Perfection* (LaSalle: Open Court Press, 1962), 254.

43. J.B. Cobb, *A Christian Natural Theology* (Philadelphia: Westminster Press, 1965), 63-70.

44. H. Zanstra, "Is Religion Refuted by Physics or Astronomy?" *Vistas in Astronomy* 10 (1967): 1-21.

5

Newton and Christianity[1]

Richard S. Westfall

Living as we do in the twentieth century in the shadow of Darwinism, an issue that remains very much with us, we tend automatically to think of the question of science and religion in terms of conflict. The titles of books, such as Andrew D. White's *History of the Warfare of Science with Theology*, serve powerfully to reinforce our tendency. The seventeenth century, in contrast, saw the matter differently. To be sure, there were a few who, even then, thought in terms of a conflict between science and Christianity; on the one hand, I have Thomas Hobbes in mind, and, on the other, theological conservatives. To me, at least, such men appear to have stood decidedly on the fringes of opinion, while the majority, especially the majority of scientists, and more especially still the majority of scientists in Britain, saw rather harmony between the two realms of knowledge and activity. The works of God magnify the glory of God. They had studied the works. Glory they had found, and from one end of the century to the other they raised their hymn of praise to its author. Robert Boyle, a most prolific author, sounds the theme in virtually every one of his many books, and toward the end of his life, in case the message had failed to be heard, he formulates his definitive statement of it in *The Christian Virtuoso*, a title that we can translate without serious inaccuracy into more familiar language as *The Christian Scientist*. John Ray, the leading naturalist of the age, states the same idea in his book, *The Wisdom of*

God Manifested in the Works of the Creation. There were so many other versions of a similar outlook that it would be difficult to list all of them.

The theme was prominent also in the works of Isaac Newton. "When I wrote my treatise about our Systeme," he begins a letter to the theologian Richard Bentley, "I had an eye upon such Principles as might work wth considering men for the beleife of a Deity...."[2] From such a point of view, science appears as the handmaiden of religion. Its purpose is not to manipulate nature for the material benefit of mankind, but to demonstrate the existence of the Creator. The "main Business of natural Philosophy," Newton asserts elsewhere (and let me insist on the significance of the adjective "main"), "is to argue from Phaenomena without feigning Hypotheses, and to deduce Causes from Effects, till we come to the very first Cause, which certainly is not mechanical...." Every step along the way brings us closer to the goal "and on that account is to be highly valued."[3] Again I want to insist on the significance of the phrase, "on that account."

John Ray, the naturalist, found God in the multiplicity of nature and the perfect adaptation of every creature to the life it must live. Newton the physicist, in contrast, found him in the structure of the cosmos. Let us assume, he argues in the letter to Bentley quoted above, that in the beginning matter was evenly diffused through the universe. If the universe had been finite, mutual gravitation would have caused all matter to congregate together in a single body. If the universe had been infinite, matter would have congregated in a number of bodies held in equilibrium by their mutual attractions, and if all matter had been luminous, we could explain the system of the sun and stars by natural causes alone. But all matter is not luminous, and by the operation of natural causes alone we cannot explain the separation of luminous from nonluminous matter. For this we have to call upon an intelligent Creator. Moreover, the sun is at the center of our system. If chance, that is, the blind operation of natural causes, had formed our system, the central body could just as well have been without light and heat; it required a God to understand the necessity that the central body be a sun. Why is there only one sun in our system? God saw that one was sufficient. Moreover, all of the planets move in the same direction in the same plane, an ordered whole, as we plainly see when we compare the planetary system with the disorder of comets.

> To make this systeme therefore wth all its motions, required a Cause wch understood & compared together the quantities of matter in ye several bodies of ye Sun & Planets & ye gravitating powers resulting from thence, the several distances of the primary Planets from ye Sun &

secondary ones from Saturn Jupiter & ye earth, & ye velocities wth wch these Planets could revolve at those distances about those quantities of matter in ye central bodies. And to compare & adjust all these things together in so great a variety of bodies argues that cause to be not blind & fortuitous, but very well skilled in Mechanicks & Geometry.[4]

I wish to contend that these arguments represent the traditional element in Newton's religion. It is the most visible part of his religion, and until recently it was virtually the only aspect of it that was known. If we concentrate on these arguments alone, they are apt to mislead us, for two different reasons. In the first place, we may make the mistake of taking the arguments at face value. But Newton did not find God in nature. Quite the contrary, he imposed God upon nature. That is, the arguments did not so much derive from the study of nature as descend from the long tradition of Christianity in western Europe. Consider the letter to Bentley that I cited. If God created an ordered universe, whence arose the disorder of the comets, to which Newton himself referred? By what criterion did Newton determine that the order of the planetary system was more typical of the whole than the disorder of the comets? If it was the purpose of God to reveal his wisdom by making the planets move in the same plane, we need to note that he bungled egregiously. The planetary planes are inclined to each other by as much as five degrees. What eighteenth-century artisan, building an orrery, would have tolerated an error that large? That is, Newton's arguments reveal above all a determination to find God in nature. They are the deposit of centuries of Christianity in the West, an inherited piety, that part of Newton's religion not yet disturbed by the rise of modern science.

The second reason why Newton's arguments for the existence of God are apt to mislead us is that we may mistake his manifest piety for the whole of his religion. In fact, the story was far more complicated. It was more complicated with all of the English scientists. When we read only one or two of their refutations of atheism, we may find them impressive testimony, but by the time we read the tenth repetition of the same argument, we begin to sense some uneasiness behind it. Boyle, for example, appears to have been aware that the ground was shifting under the traditional foundations of Christianity.

Newton was also aware of that fact, but his reaction differs from Boyle's. Instead of trying to shore up the established foundations, Newton questions orthodox theology and rejects some of its teachings that he found contrary to reason. In order to discuss this issue satisfactorily, we must make a distinction between religion and theology. Nothing that I may say is meant to question Newton's sincerity in the passages quoted above. There can be no doubt that Newton believed in the existence of God. Merely to state his belief in these terms is to grossly undervalue it. His belief in God was a living faith that suffused his entire life and gave it meaning. Equally he was convinced that science is in harmony with religion. The God demonstrated to exist from his work in the creation is not, however, necessarily identical to the God of received Christianity. In the case of Newton, he certainly was not.

We cannot effectively explore this aspect of Newton's religion through his published works. Newton was a man who feared controversy. In the matter of religion, he had good cause to fear controversy, for he had much to lose. While he was at Cambridge, the views that he came to hold during his stay there would have been grounds for his immediate dismissal, and after his move to the Mint in London, those same views made him ineligible, according to the law of the land, to hold a position in the government. Knowing this full well, he took care that his religious beliefs did not become matters of public knowledge. Newton was also a compulsive writer, however, and he left behind a huge collection of private papers. Only recently have the great bulk of his theological manuscripts become available for study, and they have a story to tell that differs from the one we find in his published works. (Note that after the auction in 1936 of the papers owned by the Portsmouth family, the large majority of theological manuscripts passed into private possession and were ultimately willed to the Jewish National and University Library in Jerusalem, where they became available less than two decades ago.) Since the papers remain largely unknown, let me cast my account in the form of a description of their content with an implicit interpretation embedded in the description.

The papers reveal that Newton began the serious study of theology about 1670, when he was approaching the age of thirty. Note that Newton began intensive theological study when he was a young man at the very height of his powers. Contrary to what has sometimes been asserted, such study was not solely the occupation of his old age, although he did devote a great deal of time to it then as well. Although there are a few dates in the theological manuscripts, one has to rely primarily upon the evidence of Newton's hand in placing them chronologically. In a previous article,[5] I included a checklist of the theological manuscripts with my considered opinion as to their dates. I shall draw upon that scheme of dating in the present account without attempting to repeat it all here.

We have every reason to think that Newton was entirely orthodox when he turned to theology. In the recent past he had taken solemn oaths to that effect on three separate occasions. With each of his two degrees, in 1665 and 1668, he had sworn to three articles, one of which affirmed the faith of the Anglican Church, and when he had accepted a fellowship in College of the Holy and Undivided Trinity, he had sworn that he would uphold the one true religion, in a context that equated the one true religion with the doctrine of the Anglican Church. Although he had not taken an oath with the Lucasian Chair, it did embody a similar requirement; its statutes made heretical belief grounds for dismissal. Everything we know about Newton indicates that he would not have sworn falsely. He must have been orthodox at the time.

In accordance with his universal practice when he turned to a new topic of study, Newton began a systematic compilation of his reading notes, which would organize his new knowledge in an effective manner. He purchased a notebook in which he entered a series of headings under which he expected to gather the fruits of his study.[6] The pages under some of the headings—"The Miracles of Christ," for example—remained blank. Such was not the case for "God the Son" and "Concerning the Trinity," which immediately became the foci of his interest. Initially, the Bible furnished his reading, and the intensive study of the Scriptures at this time furnished the foundation of the detailed knowledge of them that he carried through his life. From the Bible he advanced to the early Fathers of the Church. It is necessary to plunge into the manuscripts themselves in order fully to appreciate them. They reveal a vast program of reading that led Newton through all of the important Fathers of Christianity. They reveal as well a driving passion, which animated the study. What we meet in the manuscripts is not conventional piety assembling arguments designed to defend the tradition. Quite the contrary, what we meet is the passion of a rebel who had convinced himself that the received tradition was mistaken. "Mistaken" is far too mild a word. Newton had convinced himself that the received tradition was a fraud perpetrated by evil men in the fourth century who, for their own selfish purposes, had wilfully corrupted the entire heritage. Newton's determination to unmask this ancient crime, together with his study of alchemy, absorbed virtually all of his time for the following fifteen years before a visit from Edmond Halley started the investigation that resulted in the *Principia* and altered the tenor of his existence.

To be specific, Newton embraced Arianism, a heresy of the fourth century which had struggled with Trinitarianism (as expounded especially by Athanasius) for the soul of Christianity. Arianism is similar to modern Unitarianism insofar as it rejects the full divinity of Christ. It differs from Unitarianism in its belief that Christ was also not wholly human. The Arian Christ was a created intermediary between

God and humanity. Newton hints broadly at an Arian Christology in his theological notebook. In later manuscripts he does more than hint. Among them there is what appears to me as his most explicit statement, a sheet from the period 1672-1675 with twelve propositions on the nature of Christ which are wholly Arian in tone.[7] From the position taken in the early seventies Newton never retreated, although in one manuscript treatise he does appear to have gone further yet. In his old age, he was still composing Arian statements on the nature of Christ.

Along with theology, Newton developed an early interest in the prophecies. He composed his first interpretation of the Book of Revelation in the early seventies, and he worked assiduously at expanding and revising it through the rest of the decade and into the early eighties. His interest in the prophecies is well known. After his death, his heirs published his *Observations upon the Prophecies*. A work of surpassing tedium, which all but the tiniest handful have been spared the necessity to read, the *Observations* defies the reader to find a point in their meandering discussion. The published manuscript was a product of Newton's old age deliberately made obscure in order to conceal his point. The reader of his papers has no problem in finding a point in his early interpretation. It embodies an adaptation of the standard Puritan interpretation, which hinges on the concept of the Great Apostasy. To Puritan exegetes, the Great Apostasy was Roman Catholicism. To Newton, the Great Apostasy was Trinitarianism. That is, his interpretation of the prophecies offers an alternative statement of his theological position. The plagues and the vials of wrath in the Book of Revelation were prophetic forecasts of the barbarian invasions of Europe, God's punishments on a stiff-necked people who had gone whoring after false gods.

The work contains an inner tension. On the one hand, it implies a chronology. In Newton's interpretation Revelation contains repeated references under different figures to a period of 1260 years immediately preceding the sounding of the seventh trumpet, which will announce the second coming of Christ and the Final Judgment. Newton was quite explicit in dating the beginning of that period in 607, from which it follows that he placed the Final Judgment nearly two hundred years in the future. Not by any standard would one call that imminent. On the other hand, there is a sense of expectation in the work that is impossible to miss. The meaning of the prophecies is finally being revealed, and the Great Day must correspondingly be at hand. Intense passion accompanied the inner tension. "Idolators, Blasphemers & spiritual fornicators," Newton thunders at the Trinitarians in the silence of his study in Trinity. And, because only Scripture could adequately convey his fury, "Seducers waxing worse and worse, deceiving and being deceived—such as will not endure sound doctrine but after their own lusts heap to themselves teachers, having itching ears and turning away their ears from the truth unto fables."[8] There is no doubt in my mind that Newton considered himself as one of the remnant persecuted by the dragon, one of the saints on whom the Beast made war, and he appears to have been expecting early release. Once again, the Great Day must be at hand. Against whom was the passion directed? There is also no doubt in my mind that it was directed at flaccid orthodox Cambridge all about him, the Cambridge from which Newton had largely withdrawn as he isolated himself within the fastness of his own study.

The interpretation of the prophecies implies another program of study, indeed two more. In order to grasp their message, Newton needed to assure himself that he had their correct text. To this purpose he collated some twenty-five different Greek versions of the Book of Revelation to establish the true text, and he combed the Bible, of which he considered the prophecies to be the central books, to find confirming passages. The interpretation also implied an exact correlation between the prophetic text and the events of history as they later transpired. Although he read and used the works of his contemporary authorities, he had no intention of resting satisfied with anything less than original sources. For example, Revelation speaks of a silence of half an hour before the sounding of the first trumpet. In Newton's scheme, half an hour corresponded to seven and a half years, which ended in this case with the war between Theodosius and Maximus. In order to track the movements of Maximus in the early 380s, Newton called upon the testimony of Zosimus, Pacatus, Sulpitius Alexander (an eyewitness quoted by Gregory of Tours), the letters of Jerome and Ambrose, the *Annalium Boiorum*, and Gothofredus's commentary on the Theodosian Code. One brief passage on the period from Constantine through 380 cited Theodoret, Sozomen, Socrates (the historian), Ammianus, Claudian, Zosimus, Eusebius, Sigonius, Jerome, Eunapius, Libanus, Marcellinus, Victor, Gregory the Presbyter, Pacatus, Symmachus, Idatius, and Cassiodorus, plus the Theodosian Code, the Alexandrian Chronicle, and the letters of Ambrose.[9] Do not mistake my meaning. No one, I believe, would confuse Newton's interpretation of the prophecies with great history. The point I wish to assert is that the interpretation does not rest on superficial research.

Repeatedly the interpretation reveals itself as the work of the man who had recently composed *De methodis*, the definitive exposition of his fluxional calculus, and would soon write the *Principia*. His intention, he insists, was to methodize the study of the prophecies. He sets forth fifteen rules of interpretation—consistently rationalizing rules, such as always to assign only one meaning to one figure, and always to prefer the simplest and most literal meaning—followed by a catalogue of seventy figures and a section called the *Proof*, which cites the evidence that supports the meanings he assigns to the seventy figures. Newton was convinced that there had been an accepted language of prophecy in the ancient world, which he could decipher. For the *Proof*, he combed the entire Bible plus the writings of Achmet, an Arabian authority on prophecies, and Artemidorus, the Hellenistic author on the interpretation of dreams, for corroborating instances. His stated goal was to free the interpretation of the prophecies from individual fancy and to reduce it to demonstration. In its original version, he cast the interpretation in Propositions, though he later altered the word to Positions.

In the early eighties, Newton's interpretation began to suppress the passion that marks the early version and to shift toward the colorless chronology of the early church that characterizes the version finally published after his death. The Great Apostasy, the concept that offers the key to the whole, becomes indistinguishable in his exposition from Roman Catholicism. No single word that I have found explains the change. I will offer the speculation that the rising spectre of James Stuart and the threat of a Catholic succession led a man who hated Catholicism passionately to seek common ground with English Protestantism and to suppress those features of a work, albeit unpublished and unknown, which would have alienated him from other non-Catholics.

In the early eighties Newton also began to compose a new theological work to which, in the least chaotic of its manuscript remains, he gave the name, *Theologiae gentilis origines philosophicae.*[10] It was easily the most important theological treatise he ever wrote—or better, worked on—and traces of its continuing presence in his thought show up through the rest of his life.

The central concept of the *Origines* asserts the existence of an original pure religion of Noah and his family, who had worshipped the one true God, the Creator of the universe. The innate tendency toward superstition and idolatry in mankind had led to the corruption of this religion and the worship in its stead of twelve gods drawn from the world of nature, their common ancestors deified, but given different names as each people appropriated them to their own history. God had made repeated efforts to lead mankind back to the true religion, sending prophets to the Jews and ultimately Jesus Christ to the Gentiles. Jesus had come to recall mankind to the true religion, which rested on two commandments, to love God and to love one's neighbor. He had added nothing to that religion, Newton insisted; he had come only to call mankind back to its original worship.

We must be careful not to misinterpret the *Origines* and to miss the radical thrust that the seemingly quaint themes of Noah and his children conceals. To Newton, Noah was merely the name most familiar to readers in Christian Europe; Egypt was more basic than Israel. He treats the historical books of the Old Testament as nothing more than the historical records of the Jewish people, records no more authoritative than those of other peoples to which he compared them. Above all, the *Origines* deflates the role of Christ in human history. Christ came to call mankind back to the one true religion, and to that religion he added nothing. Its two basic commandments, Newton insisted, "always have & always will be the duty of all nations & the coming of Jesus Christ has made no alteration in them."[11] That is, Christ did not signal a new dispensation and a new religion; he merely recalled mankind to an old one. Trinitarianism, when in its turn it corrupted the restored worship once more, repeated the idolatry of earlier ages by worshipping a man as god. Along with its deflation of Christ, the *Origines* likewise suggests that the one true religion is known to mankind from the study of nature. Newton finds evidence that everywhere the original inhabitants of the earth had worshipped in similar temples, *prytanea* as he calls them, similar in plan to the Jewish Temple. The Roman temples of Vesta offered perhaps the most familiar version of such temples, which embodied a representation of the universe.

> The whole heavens they recconed to be ye true & real temple of God & therefore that a Prytanaeum might deserve ye name of his Temple they framed it so as in the fittest manner to represent the whole systeme of the heavens. A point of religion then wch nothing can be more rational.[12]

The *Origines* argues that by the proper study of nature people can recognize their Creator and their duties toward him. Christ had affirmed no more.

Newton believed in the concept of divine revelation; as with so much, however, the concept carried a new meaning for him. As an Arian he did not accept that the Bible was the revelation of mysteries beyond reason unto life eternal. I have indicated that he did not consider the historical books of the Old Testament to have been divinely revealed. Prophecy was the central element of revelation— prophecy whereby God, through its fulfillment, demonstrated his dominion over history.

The *Origines*, meanwhile, appears to me to go beyond Arianism. Although it differs in its attitude from the deists' open hostility toward Christianity, its content is remarkably similar to their works, and I am prepared to argue that it must be seen as the first of the deist tracts, even if it remained unknown to any significant number and probably to everyone. As I have indicated, the *Origines* continued to appear important to Newton, and passages drawn from it appear in the Thirty-first Query to the *Opticks* and the General Scholium to the *Principia*. In his old age, he sanitized it thoroughly, and that manuscript was published after his death as the *Chronology of Ancient Kingdoms Amended*, a work that rivals the similarly sanitized interpretation of the prophecies in tedium.

The *Principia* marks a break in Newton's theological activities. There are a few theological manuscripts from the two following decades, but not many. Sometime during the first decade of the eighteenth century Newton returned to theology again, and he devoted massive amounts of time to it for the rest of his life. However, the effort was concerned entirely with reshuffling old ideas; he added nothing new. Like his career in science, his period of creativity in theology belonged entirely to his younger age.

A survey of Newton's activity in theology raises a number of questions that I, at least, cannot avoid. No one disputes the assertion that Newton was a scientist of major, indeed monumental, importance. What influence did his theology have on his science? I need to say that it is unclear to me that we can speak validly of any influence. I mean specifically his theology, not his religion. The influence of his religion on his science is, I believe, universally admitted, and I do not challenge that conclusion. His theology, by which I mean explicitly his Arianism and the associated interpretation of the prophecies, is another matter. Perhaps we can find echoes of the Arian God in the Pantocrator of the "General Scholium," but this leaves us still on such a high level of generality that it tells us very little. If we want to descend to the details of Newton's science, as it is found in the *Principia* and the *Opticks*, I am unable to trace any line of influence that has substance.

Rather I prefer to trace the influence in the other direction. Theology was the activity with the historically established role in European civilization, a role beginning to be challenged, for the first time in more than a millennium, by a newly rising enterprise—modern science. It appears to me that we are far more apt to find the lines of influence running in this direction.

There have been a series of articles during recent decades that stress the traditional elements in Newton's religion.[13] All of them, which are based on the published writings of Newton without the advantage of access to the manuscripts this present discourse draws upon, stress the role of the Bible in Newton's religious thought. There is no doubt that Newton considered the Bible to be the word of God. Nevertheless, as I have indicated, what Newton had in mind when he said "Bible" was far from identical to what that word had meant to the previous

Christian tradition. All of these discussions of Newton's religion ignore what appears to me to be the central fact, his Arianism. When I recall the role of Arianism in early Christianity and the role of its offspring, Unitarianism, in the modern world, I find it impossible to ignore the influence of science on his religion. What I have in mind when I say "science" is not what I understand as Newtonian science. To me, the concept of Newtonian science is associated indissolubly with the *Principia* and involves the critical notion of attractions. Newton assumed his characteristic theological position, however, before he had begun so much as to dream of the *Principia*. Therefore, when I speak of the influence of science on his religion, I am thinking of more basic stands associated with the scientific revolution, especially a new criterion of truth and a new locus on intellectual authority—all that Basil Willey had in mind when he spoke of "the touch of cold philosophy." The seventeenth century's touch of cold philosophy had many sources, but none of them appears to me more crucial than the rise of modern science. Like Boyle, Newton was aware that the ground under Christianity was shifting. The central thrust of his lifelong religious quest was the effort to save Christianity by purging it of irrationalities. Is this interpretation equivalent to making Newton too eighteenth-century a figure? Allow me to remind readers that Newton was not only a seventeenth-century scientist; he was also an eighteenth-century theologian who was still writing theology when Voltaire visited London.

The story of Newton and Christianity constitutes, in my perception of things, one chapter in the central drama of European civilization, the conversion of an originally Christian civilization into a scientific one. Newton was by no means alone; many others were wrestling with the same problems. Since Newton made every effort to keep his theological views private, his opinions did not enter prominently into the main channel of religious thought; insofar as they did enter, his endeavor to save Christianity by purging it only contributed to the ultimate change, which went far beyond anything he would have welcomed. How little even the greatest of us understand the consequences of our own actions! Anyone who has read the books I have written will know that I greatly admire the achievement that modern science represents. Nevertheless, I do not revel in the decline of Christianity. I am a practising Protestant Christian, and as I look about me at the chaos of contemporary civilization, especially in its American manifestation, it is far from evident to me that the change has all been gain. Nevertheless, whether I or anyone else likes it or not, it is fact, it did happen—the greatest alteration that European civilization has undergone. It is not the least product of the scientific revolution, or of Isaac Newton.

NOTES

1. Richard S. Westfall, "Newton and Christianity" © 1987 by Chair of Judeo-Christian Studies, Tulane University. Reprinted by permission of the Crossroad Publishing Company. Originally published in a slightly different form in *Religion, Science and Public Policy*, edited by Frank T. Birtel (New York: Crossroad Publishing Company, 1987).

2. Newton to Bentley, 10 December 1692; I. Newton, *The Correspondence of Isaac Newton*, edited by H.W. Turnbull, J.F.Scott, A.R. Hall and L. Tilling, 7 vols. (Cambridge: Cambridge University Press, 1959-1977), 3:233.

3. I. Newton, *Opticks*, 4th ed. (New York: Dover Publications, 1952), 369-70.

4. Newton, *Correspondence*, 3:235.

5. R.S. Westfall, "Newton's Theological Manuscripts," in *Contemporary Newtonian Research*, edited by Z. Bechler, (Dordrecht, D. Reidel Publishing Company, 1982), 129-43.

6. I. Newton, *Keynes Collection of Newton Manuscripts (MS. 2)* (Cambridge: King's College Library).

7. I. Newton, *Yahuda Collection of Newton Manuscripts (MS. 14)* (Jerusalem: Jewish National and University Library), f. 25.

8. Newton, *Yahuda MS. 1.4*, ff. 67-68.

9. Newton, *Yahuda MS. 1.4*, ff. 33-42, 55-56.

10. Newton, *Yahuda MS. 16.2*.

11. Newton, *Keynes MS. 3*, 35.

12. Newton, *Yahuda MS. 41*, ff. 6-7. This manuscript contains a version, which appears to date from the 1690s, of one chapter of the *Origines*.

13. W.H. Austin, "Isaac Newton on Science and Religion," *Journal of the History of Ideas* 31 (1970): 521-40. L. Trengrove, "Newton's Theological Views," *Annals of Science* 22 (1966): 277-94. R.S. Brooks, *The Relationships between Natural Philosophy, Natural Theology and Revealed Religion in the Thought of Newton and Their Historiographic Relevance* (Doctoral Dissertation, Northwestern University, 1976).

6

Newton's Rejection of the "Newtonian Worldview": The Role of Divine Will in Newton's Natural Philosophy[1]

Edward B. Davis

With Aristotle's laws of motion overthrown, no role remained for a Prime Mover, or for Moving Spirits. The hand of God, which once kept the heavenly bodies in their orbits, had been replaced by universal gravitation. Miracles had no place in a system whose workings were automatic and unvarying. Governed by precise mathematical and mechanical laws, Newton's universe seemed capable of running itself.[2]

That Things could not be at first produced by Mechanism, is expressly allowed: And, when this is once granted; why, after That, so great Concern should be shown, to exclude God's actual Government of the World, and to allow his Providence to act no further than barely in concurring (as the Phrase is) to let all Things do only what they would do of themselves by mere Mechanism; and why it should be thought that God is under any Obligation or Confinement either in Nature or Wisdom, never to bring about any thing in the Universe, but what is possible for a corporeal Machine to accomplish by mere mechanick Laws, after it is once set a going; I can in no way conceive.[3]

In much traditional historiography, the relationship between science and Christianity has been described in terms of conflict, with Galileo's encounter with the Holy See serving as the paradigmatic example: reason versus authority, progressive science overcoming obscurantist theology. In recent years however historians in growing number have discarded the rhetoric of confrontation when describing the relationship between religion and science. Metaphors of interaction have begun to replace the language of the battlefield.[4] This new direction in historiography is wholly appropriate for understanding Isaac Newton (1642-1727), who did not practice the radical separation of science from theology that has come to characterize the modern world, and in terms of which he himself is so often depicted. Indeed the typical textbook for a course in Western Civilization—supposing that there is such a thing—presents Isaac Newton as the grand synthesizer of terrestrial and celestial motion whose reduction of the physical universe to a concise set of mathematical laws set the stage for Enlightenment *philosophes* to remove God wholly from the present order of things.[5]

Although Newton is described in some texts as a deeply theological person, the method he used in his science and the successes it attained are seen as the epitome of the triumph of a new rationalism over an older, essentially medieval, theocentric worldview. Sometimes it is even noted that Newton personally played no part in creating the "Newtonian worldview" of the Enlightenment secularists. Whether or not this point is made explicitly, students are all too often left with the impression that Newton was himself an Enlightenment man whose theological views were incidental to his science, or even held in contradiction to it.[6] If in our own teaching we also ignore or gloss over the vast theological gulf between Newton and the philosophers who reinterpreted his physics, we encourage the very opinion the Enlightenment deists wanted us to share: that theology and modern science are fundamentally at odds. To correct this, it is not enough merely to mention (as some texts do) that Newton was a religious man who saw his work as a fundamental contribution to the argument from design; or simply to state that he wrote much more about biblical prophecy and church history than he did about either physics or mathematics. What is needed is a fresh interpretation of the central role of theology within Newton's own science, which lies at the foundation of modern science.[7] Only then will we be able to fill in the nuances that will allow us to do justice to the complex man that was Newton, let alone the complex relationship of science and theology both in his day and subsequently.

It is of course impossible for me to give even a reasonably complete account of Newton's theology in the space allotted to me here. His private theological writings contain about three million words devoted primarily to prophecy, sacred history and doctrine. Though far less extensive, his public utterances leave no doubt that natural theology also received its fair share of attention.[8] Yet in spite of substantial recent study, the relationship between Newton's public scientific life and his private religious life remains, like almost all facets of this complex man, enigmatic.[9] I do not propose completely to clarify that relationship. Rather, it is my intention to focus on one aspect of Newton's theological thought, his emphasis on the dominion of a free and powerful God, in order to show how his concept of God underlay his rejection of the rationalistic approach to natural philosophy advocated by René Descartes and Gottfried Leibniz—the very sort of natural philosophy that was, ironically, to become so popular with Enlightenment writers.[10]

DOMINION IN NEWTON'S THEOLOGY

The notion of dominion lay at the heart of Newton's theology. It is "on account of his dominion," we read in the famous General Scholium to the second (1713) edition of the *Principia*, that God "is wont to be called 'Lord God' PANTOKRATOR, or 'Universal Ruler'...." "The Supreme God," Newton continues, "is a Being eternal, infinite, absolutely perfect, but a being, however perfect, without dominion, cannot be said to be 'Lord God'...."[11] Indeed, for Newton divine perfection was virtually equated with dominion, which he understood to be manifest in the constant activity of the divine will. The highest idea of a perfect entity, he writes in a private manuscript,

> is that it should be one substance, simple, indivisible, living and life-giving, always everywhere of necessity existing, in the highest degree understanding all things, freely willing good things; by his will effecting things possible; communicating as far as is possible his own similitude to the more noble effects; containing all things in himself as their principle and location; decreeing and ruling all things by means of his substantial presence (as the thinking part of a man perceives the appearances of things brought into the brain and thence rules its own body); and constantly co-operating with all things according to accurate laws, as being the foundation and cause of the whole of nature, except where it is good to act otherwise.[12]

Hence, as he writes in another manuscript, "The wisest of beings required of us to be celebrated not so much for his essence as for his actions, the creating, preserving, and governing of all things according to his good will and pleasure."[13]

Hand in hand with Newton's belief in the dominion of God was his suspicion of pure reason. His Arian Christ was the Christ of unadorned Scripture, the Christ whom God had revealed to men, not the Christ of idolatrous reason. Newton was an Arian primarily because he did not think the Bible teaches the Trinity, not because reason told him the Bible was wrong. This interpretation differs from the widely received view of R.S. Westfall, according to which Newton could not believe in the Trinity because he found it irrational and "physically impossible."[14]

In fact Newton carefully limited the role of reason in religion. Although he certainly thought that religion should be made as reasonable as possible, a purely rational religion is something Newton never sought, claiming that it is "contrary to God's purposes that the truth of his religion should be as obvious and perspicuous to all men as a mathematical demonstration."[15] It is not enough, Newton believes, to say that an article of faith could be *deduced* from Scripture: "It must be exprest in the very form of sound words in which it was delivered by the Apostles," for men are apt to "run into partings about deductions. All the old Heresies lay in deductions; the true faith was in the text."[16] The identical opinion had been expressed in 1650 by the prominent English physician Walter Charleton, an early advocate of Gassendi's atomist variety of the mechanical philosophy. In the "Prolegomena to the Candid & Ingenuous Reader" attached to his translation of three "paradoxes" by Van Helmont, Charleton writes that "the unconstant, variable and seductive imposture of Reason, hath been the only unhappy Cause, to which Religion doth owe all those wise, irreconcilable and numerous rents and schisms."[17] As the root of religious strife, reason was really a negative, not a positive, guide to belief.

Given his emphasis on divine dominion as revealed in God's actions rather than his essence, it is hardly surprising that Newton preferred the teleological argument to the ontological. "The dominion or Deity of God," he writes in a draft of the General Scholium, "is best demonstrated not from abstract ideas but from phenomena, by their final causes."[18] Unlike the rationalists, who sought to prove the existence of God from arguments about the necessity of his being or the force of innate ideas, Newton looked to the clear evidence of his willful actions in nature, evidence to which Newton believed he had contributed in no small measure. "When I wrote my treatise about our Systeme," he tells Richard Bentley in 1692, "I had an eye upon such Principles as might work with considering men for the beliefe of a Deity & nothing can rejoyce me more than to find it usefull for that purpose."[19] The only possible cause of the frame of the world and the diversity of creatures was the will of a sovereign God. The six planets, Newton observes, all revolve about the sun in concentric circles in the same direction and almost in the same plane; the ten moons show a similar regularity. Though their orbits might continue "by the mere laws of gravity, yet they could by no means have at first derived the regular position of the orbits themselves from those laws." It is inconceivable that "mere mechanical causes could give birth to so many regular motions...." Such a beautiful system "could only proceed from the counsel and dominion of an intelligent and powerful Being."[20]

If regularity pointed to choice rather than chance, variety pointed to will rather than necessity. "Blind metaphysical necessity," argues Newton in the General Scholium, "could produce no variety of things. All that diversity of natural things which we find suited to different times and places could arise from nothing but the will of a Being necessarily existing."[21] Thus Newton finds it "unphilosophical to seek for any other Origin of the World, or to pretend that it might arise out of a Chaos by the mere laws of Nature"[22]—as Descartes had done in several places, most notably in his posthumously published treatise *Le Monde*.[23] Descartes had also banished final causes from natural philosophy, a position Newton likewise rejected with vigor. Indeed Newton writes in Query 28 of the *Opticks* that

> The main Business of natural Philosophy is to argue from Phaenomena without feigning Hypotheses, and to deduce Causes from Effects, till we come to the very first Cause, which certainly is not mechanical; and not only to unfold the Mechanism of the World, but chiefly to resolve... [ultimate] Questions.[24]

DOMINION IN NEWTON'S NATURAL PHILOSOPHY

If Newton's understanding of God's dominion shaped the theological perspective in which he placed his science, it also affected the actual content of his science, leading him to reject what he viewed as the materialism of both his predecessor René Descartes and his contemporary Gottfried Leibniz. Neither one, as Newton saw it, allowed God to exercise dominion over the creation he had made. In the unpublished treatise *De Gravitatione et Equipondo Fluidorum*, which was probably written around 1684,[25] just as he was beginning to draft the *Principia*, Newton spells out his objections to the Cartesian concept of matter, which he regarded as a path to atheism. Unable to separate the concept of matter from that of space in his mind, Descartes had concluded that matter and extension were necessarily indistinguishable. His universe was therefore a plenum; all motion took

place in closed loops, and all changes in motion were caused by direct
contact, not by forces acting at a distance. Arguing against Descartes,
Newton claims that matter "does not exist necessarily but by divine
will." Our notion of it was therefore uncertain, "because it is hardly
given to us to know the limits of the divine power, that is to say
whether matter could be created in one way only, or whether there are
several ways by which different beings similar to bodies could be
produced."[26]

Newton went on to propose a thought experiment in which he
appealed to God's power to create pieces of empty space containing no
matter, and then to move them around as if they were pieces of matter.
The analogy upon which he relied was the human ability to move the
body at will, by thought alone. The same "free power of moving bodies
at will can by no means be denied to God, whose faculty of thought is
infinitely greater and more swift." By "the sole action of thinking and
willing," God could "prevent a body from penetrating any space
defined by certain limits." If by his power God should cause some part
of space to be impenetrable, to reflect light, and to resonate when
struck, it would be impossible to distinguish that space from true body.
If God should in addition transfer those properties to other parts of
space "according to certain laws, yet so that the amount and shape of
that impenetrable space are not changed," then even the property of
motion would be created.

Finally, God could cause us to perceive such a piece of space: "For
it is certain that God can stimulate our perception by his own will, and
thence apply such power to the effects of his will."[27] If the whole
world were constituted of only such spaces, Newton concludes, "it
would seem hardly any different." Thus "*we can define bodies as
determined quantities of extension which omnipresent God endows with
certain conditions.*"[28] But if such a world would not differ from the
one we know, why think of it in this way? "I have deduced a
description of this corporeal nature from our faculty of moving our
bodies," Newton adds, "so that God may appear...to have created the
world solely by the act of will, just as we move our bodies by an act
of will alone...."

The point of this voluntarist conception of matter, as Newton is not
reluctant to say, is that "we cannot postulate bodies of this kind without
at the same time supposing that God exists, and has created bodies in
empty space out of nothing...." The Cartesian identification of matter
and extension, on the other hand, is manifestly "a path to Atheism,
both because extension is not created but has existed eternally, and
because we have an absolute idea of it without any relationship to
God," which would make it "possible for us to conceive of extension
while imagining the non-existence of God."[29]

God's relation to the frame of time and space is indeed an intimate one for Newton. Because God has "a propensity to action," it concerns his glory and majesty "that he should never and nowhere be idle."[30] The omnipresent, eternal God "is more able by his Will to move the Bodies within his boundless uniform Sensorium, and thereby to form and reform the Parts of the Universe, than we are by our Will to move the Parts of our own Bodies."[31] Influenced by Henry More's Christian Neoplatonism, his own extensive alchemical investigations, and his own commitment to a voluntarist notion of divine activity, Newton rejects the brute mechanisms of traditional mechanical philosophies, infusing the inert world of matter with the activity of the divine will—either directly through the hand of God or indirectly through active principles, which give the world a structure and order that will evince providential choice rather than blind mechanical necessity. In the end, if Betty Jo Dobbs is correct, Newton assigns control over the short range forces of alchemical, electrical, and vital phenomena to Christ, leaving the cosmic force of gravitation to God himself.[32]

A number of Newton's contemporaries certainly understood the latter to have been the case.[33] According to a memorandum written by David Gregory in May 1694, Newton also gave God the responsibility of preventing the stars from collapsing together under the very attraction which he caused.[34] This is probably what Newton had had in mind fifteen months before when he had agreed with Bentley that if

all the matter were at first divided into several systems & every system by a divine power [were] constituted like ours: yet would the outward systemes descend towards the middlemost so that this frame of things could not always subsist without a divine power to conserve it.[35]

Thus in Query 31 Newton describes Nature as

very conformable to her self and very simple, performing all the great Motions of the heavenly Bodies by the Attraction of Gravity which intercedes those Bodies, and almost all the small ones of their Particles by some other attractive and repelling Powers which intercede the Particles. The *Vis inertiae* is a passive Principle by which Bodies persist in their Motion or Rest, receive Motion in proportion to the Force impressing it, and resist as much as they are resisted. By this Principle alone there never could have been any Motion in the World. Some other Principle was necessary for putting Bodies into Motion; and now they are in Motion, some other Principle is necessary for conserving the Motion.[36]

In the following pages, Newton elaborates on the inadequacies of a purely mechanical world. Without active principles, he argues, the quantity of motion in the world would decrease. What he had in mind here—that collisions are rarely elastic and that rotating vortices quickly slow down—fails to distinguish between what we now call momentum and energy. But it would not be misleading to suggest that his insight, despite serious difficulties, captured the essential thrust of the law of entropy: the universe is running down. "Seeing therefore the variety of Motion which we find in the World is always decreasing," he concludes,

> there is a necessity of conserving and recruiting it by active Principles, such as are the cause of Gravity, by which Planets and Comets keep their Motions in their Orbs, and Bodies acquire great Motion in falling; and the cause of Fermentation, by which the Heart and Blood of Animals are kept in perpetual Motion and Heat; the inward Parts of the Earth are constantly warm'd, and in some Places grow very hot; Bodies burn and shine, Mountains take fire, the Caverns of the Earth are blown up, and the Sun continues violently hot and lucid, and warms all things by his Light. For we meet with very little Motion in the World, besides what is owing [either] to these active Principles [or to the Dictates of a Will]. And if it were not for these Principles the Bodies of the Earth, Planets, Comets, Sun, and all things in them would grow cold and freeze, and become inactive Masses; and all Putrefaction, Generation, Vegetation, and Life would cease, and the Planets and Comets would not remain in their Orbs.[37]

The end of this passage, added in the 1717 edition, suggests a further, more cosmic, sense in which Newton believed the universe was running down. By virtue of their great masses, Jupiter and Saturn noticeably perturb one another's orbits and those of passing comets, which in turn perturb the rest of the planets. Eventually these perturbations would accumulate, Newton states, "till this System wants a Reformation."[38]

A few years before his death Newton confided to John Conduitt what may have been the full meaning of this cryptic remark. It is Newton's conjecture, Conduitt records, "that there was a sort of revolution in the heavenly bodies." Vapors and light from the sun "had gathered themselves by degrees into a body and then attracted more matter from the planets," at length forming a new planet and then a comet, which eventually falls into the sun and replenishes its matter. The comet of 1680, Newton thought, would someday meet the same fate, at which time "this earth would be burnt" and all animals would perish. Apparently he believed that something like this had happened previously, for the earth bears "visible marks of ruin upon it which could not be effected by a flood only." When Conduitt asks how the earth could have been repeopled if this had ever happened, Newton replies that "the power of a creator" was required.[39] It is important to realize that Newton did not see the need for direct divine action as a problem within his system. Quite the contrary. As David Kubrin has shown in this case,[40] and as I will argue below from the Leibniz-Clarke correspondence, Newton believed that God specifically intended to make a world that needed periodic reformation, in order to make his governance obvious to his creatures.

MIRACLES, MECHANISMS AND THE DOMINION OF AN EVER ACTIVE GOD

These questions about the cause of gravitation and the stability of the cosmos were fundamental to Newton's dispute with Leibniz about the nature of God's ongoing relation to the world. And once again, just as in his disagreement with Descartes, Newton's theology molded his position. Leibniz spells out his differences with Newton in a letter to Johann Bernoulli from December 1715. What Newton thinks, the German complains,

> seems plainly absurd to me, namely that the motion of the world-machine will come to cease unless from time to time restored by God. Thus miracles are necessary to him, and he will prove unable to explain his attraction without perpetual miracles.[41]

Leibniz debated these issues with Samuel Clarke, a friend and disciple of Newton's who acted as his spokesman in the debate, starting in late 1715 and ending a year later with Leibniz's death. Although Clarke was a capable theologian who could have debated Leibniz entirely on his own, surviving manuscript evidence indicates that Newton was intimately familiar with Clarke's arguments, perhaps in some cases even suggesting them himself. There can be little doubt that Newton endorsed what Clarke wrote.[42]

Leibniz opens his attack on Newton's views by questioning Newton's belief that the world is running down. If God has "to wind up his Watch from Time to Time," Leibniz claims, then he lacks "sufficient Foresight to make it a perpetual Motion." The Newtonians, on the other hand, obliged God "to clean it now and then by an extraordinary Concourse, and even to mend it, as a Clockmaker mends his Work...." Against this, Leibniz holds that God works miracles not "in order to supply the Wants of Nature, but those of Grace. Whoever thinks otherwise, must have a very mean Notion of the Wisdom and Power of God."[43]

Clarke (and Newton) did not agree. God is not a watchmaker, for the world is not a watch: it is utterly incapable of running on its own. God is rather "himself the Author and continual Preserver" of the forces in the world, so that "nothing is done without his continual Government and Inspection." What follows is a remarkable passage—remarkable, that is, because in it Clarke (on behalf of Newton) explicitly rejects the clockwork metaphor that is so often associated with Newtonian science:

> The Notion of the World's being a great Machine, going on without the Interposition of God, as a Clock continues to go without the Assistance of a Clockmaker; is the Notion of Materialism and Fate, and tends, (under pretence of making God a Supra-mundane Intelligence), to exclude Providence and God's Government in reality out of the World.[44]

Just as Newton objects to the Cartesian notion of matter because it does not explicitly require a creator, for the same reason Clarke objects to the Leibnizean notion of the world machine. Casting away the clockwork metaphor, Clarke turns to the much more truly Newtonian image of the world as under the constant supervision and governance of a God who works out his perfect plan. The wisdom of God, he argues, does not involve "making Nature (as an Artificer makes a Clock) capable of going on Without him: (for that's impossible; there being no Powers of Nature independent upon God, as the Powers of Weights and Springs are independent upon Men;)...." The wisdom of God involves rather "framing Originally the perfect and complete Idea of a work, which begun and continues, according to that Original perfect Idea, by the Continual Uninterrupted Exercise of his Power and Government."[45]

Clarke had gone straight to the heart of the matter. God is no absentee landlord, a perfect watchmaker whose work never needs adjustment; he is instead an omnipotent governor who exercises his dominion directly and continually as active cause of all that comes to pass. "[W]ith regard to God," Clarke observes, there are "no Powers of Nature at all, that can do any Thing of themselves...."[46] The whole order of nature is thus a constant divine work, unfolding the original perfect design. Where Leibniz insists on limiting God to what intelligible mechanisms could accomplish, Clarke allows God the freedom to act in any way for reasons known only to him:

> For why was not God at Liberty to make a World, that should continue in its present Form as long or as short a time as he thought fit, and should then be altered (by such Changes as may be very wise and fit, and

yet Impossible perhaps to be performed by Mechanism), into whatever other Form he himself pleased?[47]

Part of Clarke's (and Newton's) view of the constant divine governance of the world is the blurring—or perhaps even the outright elimination—of the distinction between natural and supernatural events. Indeed Clarke takes precisely this step, arguing that "Natural and Supernatural are nothing at all different with regard to God, but distinctions merely in Our Conceptions of things." When the sun moves across the sky every day, we call it natural; if its motion ceases for a day, we call it supernatural. But, says Clarke, "the One is the Effect of no greater power, than the Other; nor is the One, with respect to God, more or less Natural or Supernatural than the other." This is because God is present to the world "as a Governor; Acting upon all Things, himself acted upon by nothing." Deliberately echoing Newton's words in the General Scholium, Clarke adds, "He is not far from every one of Us, for in him We (and all Things) live and move and have our Beings."[48] With regard to God, no possible thing is more miraculous than any other. Miracles are simply *unusual* acts of God, but no more or less acts of God than ordinary events. The raising of a dead human body and the sudden stopping of the earth's motion are called miracles; the ordinary generation of a human body and the continual motion of the earth are called natural, "for no other Reason, but because the Power of God effects one usually, the other unusually."[49]

This same understanding of miracles is implicit in several unpublished papers of Newton's. In one of them we read:

> For Miracles are so called not because they are the works of God but because they happen seldom & for that reason create wonder. If they should happen constantly according to certain laws imprest upon the nature of things, they would no longer be wonders or miracles, but might be considered in Philosophy as part of the Phenomena of Nature [notwithstanding their being the effects of the laws imprest upon Nature by the powers of God] notwithstanding that the cause of their causes might be unknown to us.[50]

What Newton assumes here, I contend, is that God does *all* things in nature, whether usual or unusual. Most things he does by laws he established, and these we consider natural. Some things, which happen seldom and therefore give us reason to marvel, he does without laws. The word "miracle" is reserved for unusual events simply because ordinary events do not create the same degree of wonder. Newton's point about miracles not being so called because they are divine works is thus seen to be etymological: "miracle" derives from the Latin verb *mirari*, to create wonder or astonishment. Considering Newton's beliefs about divine activity in nature, he could hardly have meant that miracles are *not* acts of God; his point can only have been that they are simply *extraordinary* acts of God.[51]

This is fully consistent with his reply to Leibniz's other charge, that he could not explain gravitation without making it a perpetual miracle. Behind Leibniz's claim is the assumption that the universe is essentially Cartesian, completely full of matter in vortical motion. On this view the planets are swept around the sun by the vast whirlpool of subtle matter that constitutes the region of space we call the solar system. Although planets have a tendency to recede from the center of the vortex (the sun) owing to their inertia, they are held in check by the press of all the other matter in the universe, which is a plenum. Thus for Descartes and Leibniz, there is (and can be) no "attraction" between the sun and the planets, pulling them out of otherwise straight paths into elliptical orbits. There can indeed be no "forces" at all; there is only matter and motion, with the direct contact of bodies as the only mechanism capable of changing motion. This is why Leibniz told Clarke that the revolution of a body about "a certain fixed Centre, without any other Creature acting upon it" is something which "could not be done without a Miracle; since it cannot be explained by the Nature of Bodies."[52]

Leibniz had first aired this complaint in a February 1711 letter to Nicolaus Hartsoeker which was published in the 5 May 1712 issue of the weekly *Memoirs of Literature*. There it was seen by Newton's disciple Roger Cotes, who called it to Newton's attention. Newton's reply took the form of a letter to the editor of the *Memoirs*. A surviving draft of the letter reveals the close link between Newton's insistence on the reality of gravitation and his acceptance of the givenness of a world created by the will of God. Gravity should not be called a miracle just because no mechanical hypothesis had been offered to explain it, Newton said. We cannot give a mechanical explanation for the hardness, inertia or extension of a body, yet we do not call these miraculous:

> They are the natural real reasonable manifest qualities of all bodies seated in them by the will of God from the beginning of creation & perfectly uncapable of being explained mechanically.... And therefore if any man should say that bodies attract one another by a power whose cause is unknown to us or by a power seated in the frame of nature by the will of God,... I know not why he should be said to introduce miracles & occult qualities & fictions into the world.... But certainly God could create Planets that should move round of themselves without any other

cause then [sic] gravity.... For gravity without a Miracle may keep the Planets in.[53]

DIVINE WILL AND NEWTONIAN NATURAL PHILOSOPHY

In the passage just quoted, Newton spells out precisely his fundamental disagreement with his archrival: we cannot assume that mechanical explanation exhausts the range of natural phenomena. And this is so because we live in a universe created by the will of God, who governs the world in any manner he wishes, not necessarily as we would. If God chose to produce gravity mechanically, then let a mechanical cause be sought; if not, the phenomenon was no less real, and no less lawlike. "Gravity must be caused by an agent acting constantly according to certain laws," Newton once told Bentley, "but whether this agent be material or immaterial is a question I have left to the consideration of my readers."[54]

The readers he had in mind were those of the *Principia*, where he refrains from discussing the actual cause of gravitation, preferring instead to focus on its reality as demonstrated from phenomena. This he does deliberately, intending his approach to be seen as the rejection of the kind of science advocated by his continental rivals, who sought to derive all of nature from a few principles arising from their own fertile imaginations. He summarizes his position in a draft (not actually sent) of a letter to Roger Cotes, the gifted mathematician who was supervising the publication of the second edition of the *Principia*:

> Experimental philosophy reduces Phaenomena to general Rules & looks upon the Rules to be general when they hold generally in Phaenomena.... Hypothetical Philosophy consists in imaginary explications of things & imaginary arguments for or against such explications, or against the arguments of Experimental Philosophers founded upon Induction. The first sort of Philosophy is followed by me, the latter too much by Cartes, Leibnitz & some others.[55]

The letter which Cotes actually received made no mention of those two gentlemen or their philosophies, but Cotes did not need to be told what he could see for himself. Natural philosophers may be reduced to three classes, he advises readers in his preface to the new edition. Some follow Aristotle and reduce the effects of bodies to natures and qualities, which is to tell us nothing. Others assume hypotheses as first principles of their speculations, forming an "ingenious romance" with little resemblance to reality. The third class pursue experimental philosophy, assuming as a principle nothing not proved by phenomena.

Cotes goes on to reveal the religious foundation of this third kind of philosophy. Undoubtedly the whole world, with all its diversity of forms and motions, "could arise from nothing but the perfectly free will of God directing and presiding over all." Flowing from this fountain, the laws of nature show "many traces indeed of the most wise contrivance, but not the least shadow of necessity. These therefore we must not seek from uncertain conjectures, but learn them from observations and experiments." Anyone presumptuous enough to think that he can learn the laws of nature from pure reason

> must either suppose that the world exists by necessity, and by the same necessity follows the laws proposed; or if the order of Nature was established by the will of God, that himself, a miserable reptile, can tell what was fittest to be done.

Or, to put it another way,

> The business of true philosophy is to derive the natures of things from causes truly existent, and to inquire after those laws on which the Great Creator actually chose, to found this most beautiful Frame of the World, not those by which he might have done the same, had he so pleased.[56]

To be sure, Newton did not write this; nor did he read what Cotes had written before it went to press—although Clarke did.[57] But Cotes says nothing which Newton has not already expressed in his long and distinguished career. It is by divine will, not rational necessity, that matter exists and possesses the properties that it does. As Newton put it once in an unpublished manuscript, "The world might have been otherwise then it is (because there may be worlds otherwise framed then this)[.] Twas therefore noe necessary but a voluntary & free determination that it should bee thus."[58] Divine will has ordered the universe and would renew it from time to time as he sees fit. Natural laws are actively imposed by that will and could differ from one part of the universe to another: by varying the proportions of matter and space, and perhaps even by varying the forces, Newton argues at the close of the *Opticks*, God is able "to vary the laws of nature and make worlds of several sorts in several parts of the universe."[59]

CONCLUDING REMARKS

Surely this is not the God of the Enlightenment. I can find no necessity in Newton's theology or in his natural philosophy, no trace of the rationalist God of Descartes and Leibniz, the God who became the absentee landlord of eighteenth-century deists. But if Newton was not an Enlightenment man—if Newton was not a Newtonian—then why do we continue to treat him as if he were?

How then should we understand Newton's place in the history of science and religion, and the propriety of using him as a symbol of the larger issue of the role of science in bringing about the secularization of the Western world? I agree with Westfall that this is a major issue, but I do not agree with him in seeing "science" as the most important factor in the secularization of the modern Western world. In his essay in this collection, Westfall states near the end that what he really means by the "influence of science on [Newton's] religion" is "all that Basil Willey had in mind when he spoke of 'the touch of cold philosophy.'"[60] Thus Westfall sees Newton the scientist as a sceptical critic of Christianity, which makes him a precursor (if I might use an unhistorical term) of Enlightenment deism.

I question both parts of this equation. Obviously I question its application to Newton, whom I see as a thoroughly theistic (not deistic), though not Trinitarian, thinker whose real views on God and the world would have shocked Voltaire or Hume, just as they shocked Leibniz. What is more, I question whether it is correct to say that "science," as Westfall defines it with Willey's help, is really the principal cause of secularization. What are we to make of the fragmentation of Christendom brought about by the Reformation, the crisis of religious authority that followed in its wake, and the freedom to question doctrinal formulations that came with it? What about the religious strife (which admittedly had powerful, more purely political overtones) that tore into every part of northern Europe, including England? Other, nonreligious factors also come immediately to mind: the rise of nation states; the emphasis placed by Renaissance humanists on individual achievement rather than mutual responsibility before God; the impact of new methods of textual criticism on a religion based on ancient documents. Is this all the impact of "science"? If not, is it somehow less important?

Finally, it has been a fundamental assumption of this paper that, contrary to what we are often told, theology and science were inextricably intertwined during the crucial years when the modern scientific worldview was being formed. At a much deeper level than the superficial disputes over scriptural interpretation that accompanied the reception of Copernican astronomy, theology exerted a subtle but significant influence on seventeenth-century science, driving thinkers such as Newton to reject what they perceived to be the presumptuous claims of Continental rationalism,[61] the very sorts of claims that would later be wrongly associated with his name.

NOTES

1. This essay is based rather closely on a chapter from my doctoral dissertation, "Creation, Contingency, and Early Modern Science: The Impact of Voluntaristic Theology on Seventeenth-Century Natural Philosophy" (Ph.D. diss., Indiana University, 1984), which was completed with the aid of a generous Dissertation Year Fellowship from the Charlotte W. Newcombe Foundation. I am grateful for the comments of Richard S. Westfall on this paper as well as on the original dissertation, and for those of my colleague William V. Trollinger and an anonymous reader. Except for the final section, which is new, this paper has been published twice before (E.B. Davis, "Newton's Rejection of the 'Newtonian World View.'" *Fides et Historia* 22 [1990]: 6-20 and reprinted in *Science and Christian Belief* 3 [1991]: 103-17.) It is reprinted once again with only minor editorial changes in *Faith and Science. Volume 3: The Role of Beliefs in the Natural Sciences*, edited by J.M. van der Meer (Lanham: The Pascal Centre for Advanced Studies in Faith and Science/University Press of America, 1996), and with permission from *Fides et Historia*.

2. T.H. Greer, *A Brief History of the Western World*, 4th ed. (New York: Harcourt Brace Jovanovich, 1982), 364.

3. From the fifth reply that Newton's disciple and go-between Samuel Clarke wrote to Gottfried Leibniz in the winter of 1715-16, which Clarke published as *A Collection of Papers, Which Passed Between the Late Learned Mr. Leibnitz, and Dr. Clarke, in the years 1715 and 1716* (London: n.p., 1717), 365. Here, as in all citations from this book (which will be called "LCC"), I have not retained the frequent use of italics found in the original. A modern reprint edition (H.G. Alexander, *The Leibniz-Clarke Correspondence*, [Manchester: University of Manchester Press, 1956]) is also available.

4. For a more extensive discussion see my essay review, E.B. Davis, "Blessed are the Peacemakers: Rewriting the History of Christianity and Science," *Perspectives on Science and Christian Faith* 40 (1988): 47-52, and the references cited there.

5. I am sympathetic with readers who may argue that I am creating a straw man: that textbooks on Western history are far too diverse to be covered by my blanket statement. When it comes to the treatment of Newton, however, there is in fact very little diversity. If he is treated at all, it is almost always in such a way as to connect him, either implicitly or explicitly, with the rationalist worldview that he rejected. For some "typical" texts, see those by Greer, quoted above; D.S. Kagan, S. Ozment, and F.M. Turner, *The Western Heritage*, 2nd ed. (New York: Macmillan Publishing Company, 1983); and E.M. Burns, R.E. Lerner, and S. Meacham, *Western Civilizations: Their History and Their Culture*, 10th ed. (New York: Norton, 1984). The text by M. Perry, M. Chase, J.R. Jacob, M.C. Jacob, and T.H. Von Laue, *Western*

Civilization: Ideas, Politics and Society, 3rd ed. (Boston: Houghton Mifflin, 1989), is a rare exception to the generally poor treatment of Newton and his science found in other standard texts.

6. Of course students are hardly the only people who have this view of Newton. Many physicists and general historians, not to mention professional historians and philosophers of science, are equally ignorant (in some cases, willfully and blissfully) of Newton's actual understanding of God's relation to the world. In no small part is this owing to Newton's own reticence to discuss his theological views in public.

7. Just such an interpretation has recently appeared, but too late to take it into account in this essay. I refer to B.J.T. Dobbs, *The Janus Faces of Genius: The Role of alchemy in Newton's Thought* (Cambridge: Cambridge University Press, 1991).

8. Modern scholarship of Newton's religion began with L.T. More, *Isaac Newton, A Biography* (New York: Charles Scribner's Sons, 1934), the first work to accept at face value the Arianism which so permeates Newton's private papers. McLachlan, hardly a disinterested party as a Unitarian himself, poorly edited a collection of unpublished papers bearing on soteriology and polity; H. McLachlan, *Isaac Newton: Theological Manuscripts* (Liverpool: University of Liverpool Press, 1950). R.S. Westfall, "Newton's Theological Manuscripts," in *Contemporary Newtonian Research*, edited by Z. Bechler (Boston: D. Reidel Publishing Company), 129-43, surveys all the accessible collections and attempts an overview of their contents. Prophecy is the focus of L. Trengove, "Newton's Theological Views," *Annals of Science* 22 (1966): 277-94. Natural religion receives special attention in R.S. Westfall, "Isaac Newton's *Theologiae gentilis origines philosophicae*," in *The Secular Mind*, edited by W.W. Wagar (New York: Holmes and Meier, 1982), 15-34.

9. Richard S. Westfall's basic thesis, that Newton was a precursor of the Enlightenment, is developed in the final chapter of R.S. Westfall, *Science and Religion in Seventeenth Century England* (New Haven: Yale University Press, 1973 [1958]); in R.S. Westfall, "Isaac Newton: Religious Rationalist or Mystic?" *Review of Religion* 22 (1958): 155-70; and in his justly praised scientific biography: R.S. Westfall, *Never at Rest, A Biography of Isaac Newton* (Cambridge: Cambridge University Press, 1980). I am inclined to agree more with F.E. Manuel, *The Religion of Isaac Newton* (Oxford: Clarendon Press, 1974), who stresses those aspects which remind us that Newton was not a modern man. Neither Westfall nor Manuel sees a significant interaction between Newton's science and his religion, apart from his pursuit of natural theology. W.H. Austin, "Isaac Newton on Science and Religion," *Journal of the History of Ideas* 31 (1970): 521-42, denies that Newton's theology derived from his science; Strong denies the reverse. Varying degrees of synthesis are attempted with varying degrees of success by K.-D. Buccholtz, *Isaac Newton als Theologe; ein Beitrag zum Gespräch zwischen Naturwissenschaft und Theologie* (Witten: Luther-Verlag, 1965); and by

D. Castillejo, *The Expanding Force in Newton's Cosmos* (Madrid: Ediciones de Arte y Bibliofilia, 1981), who discovers unity in a numerological scheme that Newton almost certainly did not believe.

10. Many scholars have recognized the importance of the divine will in Newton's natural philosophy, but few have written at length on this. Exceptions include A. Koyré, "Newton and Descartes," in *Newtonian Studies*, edited by A. Koyré (Cambridge: Harvard University Press, 1965), 53-114; J.E. McGuire, "Force, Active Principles, and Newton's Invisible Realm," *Ambix* 15 (1968): 154-208; and M. Tamny, "Newton, Creation, and Perception," *Isis* 70 (1979): 48-58.

11. I. Newton, *Mathematical Principles of Natural Philosophy (1713)*, 2nd ed., translated by A. Motte and revised by F. Cajori (Berkeley: University of California Press, 1934), 544. Reprinted by permission of the University of California Press.

12. Royal Society, Gregory MS. 245, folio 14a, translated from the Latin by McGuire, "Force, Active Principles, and Newton's Invisible Realm," 190. Compare Newton's almost identical statement (from University Library Cambridge Add. MS. 3965.13) on page 123 of J.E. McGuire, "Newton on Place, Time, and God: An Unpublished Source," *British Journal for the History of Science* 11 (1978): 114-29.

13. Yahuda MS. 21, folio 1r, quoted by Manuel, *The Religion of Isaac Newton*, 21f.

14. See Westfall, *Science and Religion*, 210-20, quoting 217. The present paper, and R.S. Westfall, "Newton and Christianity," in *Facets of Faith and Science. Volume 3: Role of Beliefs in the Natural Sciences*, edited by J.M. van der Meer (Lanham: The Pascal Centre for Advanced Studies in Faith and Science/University Press of America, 1996), focus on quite different aspects of Newton's theology and come to quite different conclusions. Westfall looks at the two formal elements of Newton's theology that he sees as the essence of that theology: his Christology (which was Arian) and his eschatology, which are shown to have worked together against a Trinitarian view of Christ. My paper, on the other hand, examines Newton's voluntarism, which is best understood as a theological attitude rather than a set of propositions or beliefs about God or Christ. I show that this attitude affected several aspects of Newton's thought: his Christology, his natural theology, and his natural philosophy; elsewhere I have extended this also to his view of prophecy.

Strictly speaking, then, our conclusions do not conflict, since we have studied different faces of this complex person. Nevertheless we differ in our overall view of Newton. Westfall sees Newton as a religious rationalist who could not believe an irrational doctrine (the Trinity), and who sought to remove from Christianity those beliefs which could not stand up to sceptical inquiry. I see Newton as a theological voluntarist whose Arianism derived partly from his suspicion of a doctrine (the Trinity) derived from reason rather than the plain words of the biblical text, and who sought the pure religion of the sons of Noah as an antidote to the follies that an idolatrous form of reason had brought upon European Christendom (this latter point I have not developed here).

15. Fragments from a treatise on Revelation, Yahuda MS. 1, folio 19r, printed as an appendix to Manuel, *The Religion of Isaac Newton*, 124.

16. Here Newton was commenting on 2 Timothy 1:13, "Hold fast to the form of sound words, which thou hast heard of me..." Yahuda MS. 15.1, folio 11r, quoted by Manuel, *The Religion of Isaac Newton*, 54f.

17. Quoted by J. Henry, "Henry More Versus Robert Boyle: The Spirit of Nature and the Nature of Providence," in *Henry More (1614-1687)*, edited by S. Hutton (Dordrecht: Kluwer Academic Publishers, 1990), 58.

18. A.R. Hall, and M.B. Hall, eds. and trans., *Unpublished Papers of Isaac Newton* (Cambridge: Cambridge University Press, 1962), 363.

19. Letter of 10 December 1692, in I. Newton, *The Correspondence of Isaac Newton*, edited by H.W. Turnbull et al., 7 vols. (Cambridge: Cambridge University Press, 1959-1977), 3:233. Reprinted with permission from the President and Council of the Royal Society. All future references to this set will be given as *Corres*.

20. Newton, *Mathematical Principles*, 543f. For similar comments to Bentley, see *Corres*, 3:235 and 254f. Cf. the drafts of the General Scholium in Hall and Hall, 353, 360, and 362f.

21. Newton, *Mathematical Principles*, 546. This passage was added in the third (1726) edition. Cotes expressed a similar view in his preface to the second edition (xxxi-xxxii). Cf. the selection from University Library Cambridge Add. MS. 3965.13, McGuire, "Newton on Place, Time, and God," 123.

22. I. Newton, From Query 31 in the fourth edition of the *Opticks* (New York: Dover Publications, 1952), 402.

23. See, for example, chapter 6 of R. Descartes, (1664) *Le Monde, ou Traité de la Lumière*, translated by M. S. Mahoney (New York: Abaris Books, 1664, 1979).

24. Newton, *Opticks*, 369.

25. The generally accepted date for its composition (ca. 1670) is probably wrong, as the late Betty Jo Dobbs kindly shared with me in a letter some years ago. She has made public her reasons in 139f of Dobbs (B.J.T. Dobbs, *The Janus Faces of Genius: The Role of alchemy in Newton's Thought* [Cambridge: Cambridge University Press, 1992]). The Latin text of *De Gravitatione* is printed with an English translation in Hall and Hall, 89-156.

26. Hall and Hall, 138.

27. Hall and Hall, 138f.

28. Hall and Hall, 139f, emphasis his.

29. Hall and Hall, 141-43. Cf. 132 and 145. This argument is very similar to one used by Robert Boyle, who compared God's ability to move the bodies in his world to a man's ability to move his own shadow by an act of pure volition. See his essay "Upon the Sight of One's Shadow Cast upon the Face of a River," one of the *Occasional Reflections*, in R. Boyle, *The Works of the Honourable Robert Boyle*, edited by T. Birch, 6 vols. (London: Millar, 1772), 2:401-6, where Boyle styles the world as "God's shadow."

30. University Library Cambridge Add. MS. 3965.13, folio 541v, quoted by McGuire, "Newton's Invisible Realm," 201.

31. Query 31 in the *Opticks*, 403. Cf. Query 28, *Opticks*, 370. On Newton's belief in space as the literal sensorium of God, a position which he later tried to hide, see A. Koyré and I.B. Cohen, "The Case of the Missing *Tanquam*: Leibniz, Newton and Clarke," *Isis* 52 (1961): 555-66; and Westfall, *Never at Rest*, 646-48.

32. See B.J.T. Dobbs, "Newton's Alchemy and His Theory of Matter," *Isis* 73 (1982): 511-28. Her study of Newton's alchemy (B.J.T. Dobbs, *The Foundations of Newton's Alchemy: or, "The Hunting of the Greene Lyon"* [Cambridge: Cambridge University Press, 1975]) has already become a classic. Other outstanding contributions to this aspect of Newtonian thought are McGuire, "Newton's Invisible Realm," and E. McMullin, *Newton and Matter and Activity* (Notre Dame: University of Notre Dame Press, 1978), a model of clarity.

33. This would include Locke, Wren, Gregory and Whiston. See Westfall, *Never at Rest*, 510 and 647, and R.S. Westfall, *Force in Newton's Physics*, (New York: American Elsevier, 1971), 395-400. Leibniz, another who understood this, will be discussed below. Newton's support for an ether as the cause of gravitation at the end of his life, as seen in Queries 17-24, must not be mistaken for a return to traditional mechanical explanation; Westfall, *Never at Rest*, 794. For a superb account of Newton's changing views on the cause of gravitation and other forces, see McGuire, "Newton's Invisible Realm."

34. *Corres*, 3:336.

35. Letter of 25 February 1693 (*Corres*, 255). In Query 28 of the *Opticks*, 369, published first in 1706, Newton implies that God is "what hinders the fix'd stars from falling upon one another." In the General Scholium, however, we find only the following phrase, added to the third (1726) edition: "and lest the systems of the fixed stars should, by their gravity, fall on each other, he [God] hath placed those systems at immense distances from one another." (*Mathematical Principles*, 544). Apparently the continuous action of God was no longer thought to be required for this.

36. Newton, *Opticks*, 397.

37. Newton, *Opticks*, 399f. The words in brackets were deleted from the 1717 edition, in which the last sentence was added.

38. Newton, *Opticks*, 402.

39. King's College, Conduitt Papers, Keynes MS. 130.11, quoted by Castillejo, *The Expanding Force*, 95-97. According to Gregory's memorandum of May, 1694 (*Corres*, 3:336), Newton believed that "The Satellites of Jupiter and Saturn can take the places of the Earth, Venus, Mars, if they are destroyed, and be held in reserve for a new Creation"; cf. Newton's letter to Bentley of 25 February 1693 (*Corres*, 3:253). Whether Newton believed in pre-Adamite men, I do not know and do not care to speculate.

40. See D. Kubrin, "Newton and the Cyclical Cosmos: Providence and the Mechanical Philosophy," *Journal of the History of Ideas* 28 (1967): 325-46.

41. *Corres*, 4:261.

42. The relevant evidence, which is quite extensive, is evaluated in A. Koyré, and I.B. Cohen, "Newton and the Leibniz-Clarke Correspondence with Notes on Newton, Conti, and Des Maizeau," *Archives Internationales d'Histoire des Sciences* 15 (1962): 63-126. In an earlier work, *From the Closed World to the Infinite Universe*, (Baltimore: Johns Hopkins University Press, 1957), A. Koyré had said that he was "morally certain that Clarke communicated to Newton *both* Leibniz's letters and his own replies to them" (301). In the article just cited, Koyré and Cohen observe that from their "study of the Newtonian manuscripts, the 'moral' conviction has been transformed into a demonstrable one" (67). For the correspondence itself, see note 3.

43. From his first paper to Clarke (LCC, 5-7).

44. Clarke's first reply (LCC, 15).

45. Clarke's second reply (LCC, 45).

46. Clarke's second reply (LCC, 47).

47. Clarke's fifth reply (LCC, 347).

48. Clarke's second reply (LCC, 49-53). Newton also made reference to Acts 17:27f in the General Scholium. For excellent discussions of the breakdown of the natural/supernatural distinction by Luther, Calvin and the mechanical philosophers, see K. Hutchinson, "Supernaturalism and the Mechanical Philosophy," *History of Science* 21 (1983): 297-333; and G.B. Deason, "Reformation Theology and the Mechanistic Conception of Nature," in *God and Nature*, edited by D.C. Lindberg and R.L. Numbers (Berkeley: University of California Press, 1986), 167-91.

49. Clarke's fifth reply (LCC, 351).

50. From an uncatalogued MS at Lehigh University, quoted with permission. The words in brackets are crossed out. For similar statements, see the various manuscripts printed in Koyré and Cohen, "Newton and the Leibniz-Clarke Correspondence," 72-75 and 108-15.

51. This is another place where my interpretation differs from the standard view, as seen for example in R.S. Westfall, *Science and Religion*, 204. Though Westfall grants the possibility that the interpretation I defend here could be correct, he sees the passage just cited as evidence for the confusing conclusion that "Newton both believed in and did not believe in miracles." In my view, the confusion is Westfall's, not Newton's, and arises from overlooking the voluntarist elements in Newton's thinking, which make his belief in miracles just one part of a coherent, larger theology of nature.

52. Leibniz's third paper (LCC, 69-71).

53. University Library Cambridge Add. MS. 3968.17, folio 257, as printed in *Corres*, 5:300.

54. Letter of 25 February 1693 (*Corres*, 3:254).

55. University Library Cambridge, Add. MS 3984.14, folio 1, as printed in *Corres*, 5:398f.

56. Newton (1934), pages xxxii and xxvii, respectively. John Locke had the same view of Newton's science. See A. Koyré, "Gravity an Essential Property of Matter?" in *Newtonian Studies*, edited by A. Koyré (Cambridge: Harvard University Press, 1965), 154f.

57. See Westfall, *Never at Rest*, 749.

58. Quoted with permission from MS 1031B (formerly: Burndy 16), folio 4v, Dibner Library, Special Collections Branch, Smithsonian Institution Libraries, Washington, D.C. 20560. Dobbs kindly called my attention to this passage.

59. Query 31 (Newton, *Opticks*, 403f).

60. R.S. Westfall, "Newton and Christianity," in *Facets of Faith and Science. Volume 3: Role of Beliefs in the Natural Sciences*, edited by J.M. van der Meer (Lanham: The Pascal Centre for Advanced Studies in Faith and Science/University Press of America, 1996.

61. Robert Boyle also comes to mind. See my dissertation, "Creation, Contingency, and Early Modern Science: The Impact of Voluntaristic Theology on Seventeenth-Century Natural Philosophy" (Ph.D. diss., Indiana University, 1984), for details.

7

Physical Laws as Knowledge and Belief

C.I.J.M. Stuart
and
Tom Settle[1]

INTRODUCTION

The principal aim of this paper is to argue that proper attention to the grammar and logic of scientific theories reveals a metaphysical implication: *science implies the existence of a primitive reality deeper than the empirical reality scientists regularly encounter in their work.* This primitive reality is independent of the thought and work of human beings, including scientists. I illustrate my argument using quantum theory and thermodynamics, where what I argue for helps to dissolve some paradoxes or oddities which resist more superficial analysis.

A second, related aim is to reveal some analogy between science and religion connected with how belief and knowledge function in those seemingly quite dissimilar domains. How I use the terms "belief" and "knowledge" may strike some readers as slightly idiosyncratic, though I think I am using them in a manner common among natural scientists, and I hope their meaning will be clear in context.

For the past several years my work has been concerned with the primitive concepts of physical theory: concepts not explicitly defined but more or less taken for granted in connection with the mathematical formalism. Familiar but profound examples are *time, causality, order,* and *reality*. These can never be defined in terms of anything else, but get their meaning in several ways: partly from ordinary experience, since all these terms function in commonsense attempts to organize and understand experience; partly from how technical aspects of the concepts behave in the theoretical (scientific) systems which presuppose them; and partly from how the terms function in metaphysical or theological systems which seek to embed scientific findings in a broader and deeper context.

It is a matter of controversy where to draw the line between science and metaphysics—if indeed such a line can be drawn—and it is not my business in this paper to enter that discussion generally, but I do want to discuss the way some scientists want to understand their primitive terms. I want to argue against those who wish to distance these terms as far as possible from any whisper of metaphysics. I argue that doing so introduces paradoxes which I think can only be resolved by closing the distance and allowing metaphysical insights to enter the *content* of these terms. In brief, the effort to empty science's primitive terms of any metaphysical meaning at all causes problems, though the problems do not themselves tell us exactly which metaphysics to prefer. In the main body of the paper, I discuss just one of these terms, perhaps the most crucial: reality. But I will also draw the others I have mentioned into the discussion, while regretting that space does not allow a fuller treatment.

These and other primitives are of central concern to research on the conceptual foundations of physics and, within that field, my particular concern has been with *paradoxes* associated with the physical primitives. Most of these paradoxes have aspects that are of interest to a wider audience than specialists in foundations of physics and, for the present discussion, I have drawn upon aspects of my work related to the role metaphysical beliefs play in the content of physical theory. Everyone is familiar with the idea that physical science rests on the hypothetico-deductive mode of thought; and with *hypothesis* comes *belief*.[2] Equally familiar is the idea that the fundamental postulates of physical theory are not directly testable in themselves but only through their deductive consequences. It might therefore seem entirely gratuitous to emphasize the role belief plays in physical theory. It is, however, of singular importance that quantum mechanics flatly rejects one of the fundamental beliefs of classical physics, namely, the belief that physical theory pertains to a reality that exists with definite properties independent of human observers.

I mean a rather complicated thing in saying that such a belief in reality was part of classical physics. I do not mean that it was a *finding* of classical physics: it could not be that. The primitive reality I refer to is not immediately accessible to human observers: we deal directly in appearances, in phenomena, in elements of the empirical world. But it does seem to have been the case that early scientists by and large took it for granted that appearances pointed to a deeper reality which the laws of physics were intended to delineate. Thus the eighteenth century philosopher David Hume's arguments, to the effect that experience could not prove that theories in science were true of that deeper reality, were something of a shock. But Hume's point was accepted within a century by positivists, and that acceptance still appears to influence the thought and output of very many of my colleagues in the natural sciences. Nobody has counted how many, but neither does anybody dispute that a kind of positivist reluctance to be drawn into talking about a reality beyond the reach of phenomenal testing is prevalent among workers in quantum theory.

When I talk about knowledge in science, I respect this distinction the positivists draw, and I think of knowledge as restricted to the domain of those phenomenological laws and findings about which all scientists in a field will agree. Scientific knowledge, like commonsense knowledge, is what nobody in a position to know seriously contests. This I distinguish from belief, where one can certainly be right, but one might be on one's own.

Classical physics, as I read it, hypothesized the existence of the subject matter of its own inquiry. It hypothesized space, time and causality as real, independently of human thought or experience, even if, as Hume pointed out, it could not prove them real by experiment. This contrasts starkly with the dominant interpretation of quantum theory, put forward in its early days by Niels Bohr, and promoted by him from his laboratory in Copenhagen, which for years was a Mecca for students of quantum theory.[3] On the Copenhagen interpretation, as it is called, the mathematical formalism of quantum mechanics simply provides rules for calculating experimental expectation values. The point is that, unlike classical theory, the quantum algorithms provide no *explanation by hypothesis* for the observed values. In other words, the quantum formalism does not set forth hypothesized beliefs as to the character of actual physical mechanisms that would account for our empirical observations. There are no beliefs, on this view, just empirical knowledge.

Again, in saying that classical physics incorporated belief in a reality independent of human thought (or even existence) as a primitive concept, I am not merely saying that in the early days most scientists endorsed such a belief, while perhaps most scientists today do not. I am saying that the grammar and the logic of the standard formulations of classical physics implied it. A person would have to work against the grain of what was being said within classical physics to come out denying belief in that external reality. Positivism, in its efforts to deny any metaphysical implications to science's work is just such a working against the grain.

For my point to be clear, here, I need to distinguish what is implied by the grammar of everyday English from what is implied by the more technical constraints of logic, which sometimes are more stringent, and sometimes more lenient. Of course, the technical constraints differ according to which logic one is working with, but if we assume a standard modern logic—first order predicate calculus, say, as opposed to Aristotle's logic—then making statements about some kind of entity does not imply the existence of anything of this kind. One needs to add an explicit existence statement to what one is saying for one to be committed to there being anything of the kind. This is a novel move in this century. Consequently, in addition to my making the point that the grammar of ordinary usage indicates that classical physics implies the existence of a deeper reality, I have also to argue that problems arise when one explicitly goes against this assumption.

During the pre-modern period, where opinion was divided as to whether models of the solar system—Ptolemy's, say, or Copernicus's—were intended as descriptions of reality or merely as calculational devices aimed at saving the phenomena, it would have been customary for a person holding the latter view not to leave readers in doubt that the descriptive language in which theory is couched was not intended literally.[4] In modern times, too, a few writers have been explicit about not wanting to be taken literally, for example Pierre Duhem.[5] But the way I read classical physics, it is not attended by this disclaimer. Even though physicists of this period might not have indulged in explicit belief in an underlying reality, I have not found people dissociating themselves from such a belief in any significant numbers. The scene changes in the twentieth century with quantum physics.

But my argument does not rest with this point about what we are to take classical physicists (by and large) to have meant. I shall make the extra, and main, point that trying to do without the belief in a reality deeper than appearances as part of what science refers to, trying to manage with only the phenomena—"empirical reality"—leads to unhappy and unresolvable paradoxes. I begin my argument proper by describing how this happens within quantum theory.

AVOIDABLE PARADOXES IN QUANTUM THEORY

Concerning the question of physical realism, I have encountered deeply rooted but inconsistent dispositions of mind among contemporary physicists. On the one hand, I have encountered a belief, analogous to religious faith, in the existence of an independent reality[6] that forms the proper subject-matter of physical inquiry. On the other hand, I have found an entirely opposed view, the one promoted by positivism, that has increasingly become projected onto the whole of physical science since its popularity as an interpretation of quantum mechanics. On this second view, physics has nothing to say about a supposedly independent reality. Usually, proponents of this view adopt a tone that can properly be described as one of self-congratulation in dismissing talk about *physical reality* on the grounds that they simply do not know what it *means*; the implication is that what is not explicitly defined has no meaning and, therefore, is not worthy of discussion. You will realize that I regard this as an utterly misplaced view, since, *reality* being one of the *primitive concepts* of physical science, it is *of course never defined!*

As previously noted, we may view physical theory in either of two ways: as projecting its organizing principles onto an independent reality, or as projecting them onto our empirical experiences. In the former case, we *assume* the existence of a reality independent of human knowledge. The distinction between these two concepts of reality is central to any understanding of the Einstein-Bohr controversy over how to interpret quantum theory. On Einstein's view, the organizing principles expressed by physical laws can account for the observed regularities in actual phenomena only if those principles correspond to the action of natural causalities. In this way, Einstein referred physical laws to a primitive reality. On Bohr's view, physical laws must forgo causal explanations and, instead, function as rules of correlation that subsume actual phenomena. For Bohr, physical theory pertains solely to our knowledge about empirical reality; instead of explaining phenomena, the role of theory is to provide them with mathematical organization from which we can make successful predictions.

There seems to be no a priori compulsion for us to make the one choice or the other. But the choice we make determines how we perceive the content of physical theory. Although other people might form a different impression—there are no statistics here to guide us—it seems to me that Bohr's view of physical theory is dominant throughout the whole of contemporary physics, so it is not to be dismissed lightly. On the other hand, its being the currently dominant view is not in itself a decisive consideration as to its correctness.

Quantum mechanics is concerned with what happens when systems at the atomic and sub-atomic level *interact* with the macroscopic world represented by the laboratory measurement apparatus. Because of this interaction—which is in fact utterly unavoidable for experimentalists— the Copenhagen interpretation insists that we are not able to claim that the measurements reveal properties of the quantum system considered in isolation: quantum measurements only reveal what is generated by the measurement interaction. On Bohr's view, the essential presence of the measurement interaction means that the *physical system* under study is comprised by the object-system and the measuring apparatus *inseparably*. Consequently, the only meaning allowed for the outcome of quantum measurements is the *information or knowledge* we gain from that inseparable system.

Generalization of this point of view to the whole of physics is expressed by saying that physical laws and theories describe only the correlation of knowledge gained from observations. On that view, the notion of *physical reality* takes on a particular meaning, well expressed by Max Born: "A concept refers to a physical reality only when there is something ascertainable by measurement corresponding to it in the world of phenomena."[7]

In brief, whereas Einstein considered physical laws as laws of Nature, revealing organizing principles intrinsic to an underlying independent reality, the Copenhagen view is that physical laws have their *sole* domain of reference in empirical reality. How are we to resolve this?

We cannot solve individual physical problems directly from the laws; we have to set up the appropriate equations and enter initial conditions appropriate to the problem at hand into them. The predictions so obtained are then tested against actual measurements. Notice that although the initial conditions are essential to the derivation of actual predictions, they are nevertheless *external* to the laws and equations, and so they are on a different logical footing from the ordering principles expressed in the laws and related equations. The measurable quantities specified by the laws are entities of a kind categorically distinct from the values obtained by individual measurements. The one is conceptual and qualitative, the other is a matter of quantitative fact: to specify an entity as being measurable is not the same as specifying a measure on it. But the same relations must hold between the two kinds of entity. This yields an *isomorphism*: the relations specified by the laws project isomorphically onto the outcomes of actual measurements.

Obviously, the outcomes of these physical measurements specify our *empirical knowledge* about the actual situation. The laws are logically independent both of their *manifestations* in actual phenomena and of our knowledge thereof, yet all three are logically related: The isomorphism projects from the laws onto actual phenomena *and* onto our empirical knowledge. But the matter does not end there, for the empirical knowledge in question fully satisfies the Copenhagen criterion—given in the passage cited from Max Born—for reference to a *physical reality*. Indeed, it is this very circumstance which underlies the Copenhagen idea that physical laws describe our knowledge about phenomena. The concept of the quantum wave function will help to illustrate the general point at issue here.

On the Copenhagen view, there is no material wave involved: the wave function only represents our knowledge of the system in question, the absolute square of the wave function representing the probability for the outcome of measurements. Under the Copenhagen criterion for physical reality, the wave function cannot correspond to a physical reality prior to the measurement interaction; for there is no act of observation at this stage and, hence, we are not dealing with the world of phenomena. Formally, at this stage the wave function obeys the time-symmetric Schrödinger equation, but the measurement interaction brings about a collapse of the wave function formally introduced through the action of projection operators. The collapse introduces an irreversible, time-asymmetric, change in the situation. This change is manifested as a change in the observer's state of knowledge whereby the probabilistically defined wave function is replaced by actual knowledge given as a definite measured value in the world of actual phenomena.

According to the Copenhagen criterion for a physical reality, the wave function's collapse is real, for it corresponds to an event at which the time-symmetric equation of motion, Schrödinger's equation, gives way to an irreversible measurement outcome through which an actual phenomenon becomes recorded. There is something paradoxical about this. One may ask how we can obtain a physical reality from the collapse of something that is not itself a physical reality. The paradox is not eliminated, but only made more complex, if we adopt the view that the wave function refers to a large ensemble of similarly prepared systems.

As an element of the mathematical formalism, the wave function is viewed by the Copenhagen scheme simply as a means for obtaining physical predictions. This requires that physical laws reveal ordering principles ascribed to the only physical reality accessible to us, namely, empirical reality. Consistency then requires one to say that physical laws only describe correlations between phenomena, a point repeatedly emphasized by Heisenberg. But consistency also requires a parallel interpretation throughout physics, for otherwise we should be left with a situation in which quantum mechanics would be the only physical theory that prescinded from advancing mechanisms serving to explain actual phenomena while at the same time offering no basis upon which we could understand why this should be so.

Here, however, we run into trouble precisely over the Heisenberg uncertainty relations. As far as I can tell, the great majority of contemporary physicists, including those who uphold the Copenhagen interpretation, regard these relations as expressing a *fundamental indeterminacy as an intrinsic property of nature*. Indeed, this is the view adopted by Heisenberg himself who explicitly postulates the uncertainty relations as a law of nature in direct parallel with Einstein's postulate on the invariance of the speed of light.[8]

A lot depends upon what is meant by *nature*. If we regard nature as corresponding to the totality of all possible phenomena, the Heisenberg relations may then reasonably be said to express a law of nature as confirmed by our experiences thus far. And that perception is entirely consonant with the Copenhagen interpretation. But in terms of the Copenhagen scheme, we create a paradox if, by a law of nature, we mean a property intrinsic to an independent reality. The point is this. If understood as imposing a universal limit on actual observations, the uncertainty principle is then a statement restricted to the level of actual phenomena. But if it is understood as a limiting condition on all possible observers, there is then the implication that the condition must be independent of the presence of any observer.

Consider the question as to whether *order* was already present in the earliest events following the Big Bang. We are quite evidently dealing with a situation in which no human observers were present. This is the sense I take Stephen Hawking to have in mind when he writes, "The uncertainty principle implies that the early universe cannot have been completely uniform because there must have been some uncertainties or fluctuations in the positions and velocities of the particles."[9] Here, the fluctuations in question cannot be uncertainties imposed on observers; they can *only* be indeterminacies intrinsic to the universe itself.

Anyone who accepts that such indeterminacies are implied by the Heisenberg relations flirts with an inconsistency here: the cosmological application violates the Copenhagen interpretation of the Heisenberg relations! They cannot both refer and not refer to a reality independent of observers. I argue that consistency requires those who take a realistic view of cosmology—those who think that there actually was a universe before observers came on the scene to observe it—to drop the Copenhagen interpretation of quantum theory. Dropping that interpretation does not imply adopting one or other of the more realistic interpretations currently to hand; perhaps none of them is satisfactory. But it does imply refraining from ruling out that there is one to be found. This is consistent with the work of John Bell, who showed thirty years ago[10] that there had not yet been a valid proof of the impossibility of a realistic interpretation—as indeed how could there have been, given that David Bohm explicitly constructed a consistent realistic interpretation, complete with hidden variables?[11]

THE PREDICTION PUZZLE

The enormous predictive success of science (which is exemplified by the unparalleled measure of empirical confirmation quantum theory has received), coupled with a reluctance to affirm a reality underlying empirical experience, leads to what I call the prediction puzzle.

The basis of the puzzle lies in this. We do not want to say that a theory's predictive success is due simply to chance or to repeated miracles.[12] We want to assign the cause to something rational and ordered. At a minimum we can say that it must depend on an element of regularity or consistency in the character of our empirical experiences and, hence, in the content of our empirical knowledge; but while the objective testing of such experiences can attest to their consistency, it cannot provide the source for the consistency itself. How then are we to account for consistent predictive success through theories which assume that our empirical experiences form an internally consistent system *on their own merit*?[13]

I want to say that consistently successful prediction not only tests the *external consistency* of theories against actual observations, it also presupposes the existence of some independent and self-consistent *source of regularity* for empirical phenomena. I think successful prediction "implies"[14] that physical theory has a possible model or interpretation in this independent source and, thereby, prediction is simultaneously an indirect test of the theory's *internal consistency* also. Let me illustrate what I mean from mathematics.

In mathematics we can attest to the internal consistency of a given system by showing that it has a model or interpretation in an external system about which there is no question. This cannot be a strictly logical move since trying to supply the proof would lead to an infinite regression in pursuit of a system that could be attested as self-consistent without external appeal. In direct parallel, the situation in physical science rests on the *fundamental assumption* that there exists a physical reality which is an independent and self-consistent system. Personally, I think that the prediction puzzle makes this assumption methodologically—but not metaphysically—inescapable.

From a more formal point of view, if a physical theory exhibits predictive success, then it is possible to construct a non-empty set of ordered pairs of statements $\{(T,M)_i\}$, in which the components T_i are derived from the theory in question and the M_i from experimental findings, components with the same index agreeing in their content (up to specified tolerances of experimental error). The resulting construction yields a Cartesian product defining the *predictive relation*:

$$P = T \times M \qquad\qquad [1]$$

meaning that the theory T has an external model or interpretation in M. If we accept Einstein's view that physical theory pertains to an independent reality R, postulated to be internally consistent, we can then restate relation [1] as:

$$P = T \times R \quad \text{with } M \neq R \qquad\qquad [2]$$

On the other hand, if we adopt the Copenhagen view, then the set M in relation [1] is defined only on empirical reality, E. Accordingly, we write

$$P = T \times E \quad \text{with } E \neq R \qquad\qquad [3]$$

While relation [1] is not a necessary condition for the external consistency of T—T might have models external to itself that had no empirical manifestation—it is a necessary condition for its empirical, external consistency, since there would be no detectable model external to our humanly devised theory if we had no experimental findings. Since M is properly contained in E, it follows not only that relation [3] is likewise a necessary condition, but also that if [3] is not a sufficient condition, then neither is [1]. However, the *prediction puzzle* establishes that relation [3] is not a sufficient condition, since the consistency of E (as well as M) can only be viewed as contingent. It is the consistency of M (and E) that we want to explain. We cannot use it to attest to Ts. In brief, relation [3] provides a necessary but not sufficient condition for the externally detectable consistency of T.

On the other hand, we can sidestep the prediction puzzle if we assume that T has a putative model in R. Provided R is (as we assume) consistent, T is consistent, following the kind of argument I have referred to in mathematics. It would be tempting to say that relation [2] is thereby established as a sufficient condition for T's consistency, but it cannot be a sufficient condition for our knowing about T's consistency, since access to R must be mediated by experimental findings defined on E. Thus the formal version of my argument is to propose that relations [2] and [3] *together* constitute necessary and sufficient conditions for the detectable external consistency of physical theory.

If this solution is correct, any physical theory which fails to escape the predictive puzzle is *necessarily incomplete*, since the predictive relation for such a theory cannot provide a necessary and sufficient condition for the external consistency of the theory in question. Since the Copenhagen requirement for the predictive relation, that is, relation [3], provides a necessary but not sufficient condition for detectable external consistency, it follows that the Copenhagen scheme yields an incomplete physical theory.[15]

The foregoing analysis reveals that physical theory has a dual role or aspect. Not only does physical theory prescribe what can (and, by implication, what cannot) belong to empirical reality (E). It also defines hypothetical beliefs about primitive reality (R). In this way, the mathematical structure of physical theory shapes a *system of empirical knowledge* through the predictive relation [3] and, simultaneously, it shapes a *system of theoretical belief* through the conjectural relation [2]. The dual role is not arbitrary. It is a consequence of the prediction puzzle.

It is especially important to note that relation [2] does not necessarily yield a position either of naive realism or of classical scientific realism. Nor does it necessarily yield any of the currently standard versions of scientific realism.[16] Instead, relations [2] and [3] jointly define a form of what I call *necessary realism*. The critical difference is that whereas other forms of realism take it for granted that physical theory *describes* an independent physical reality, necessary realism says that physical theory describes the character of empirical reality and, at the same time, defines *beliefs* about the character of primitive reality.

The concept of necessary realism, which I am introducing, has some interesting implications. One of these follows from the isomorphisms associated with physical laws. Those isomorphisms do not establish that the laws have a model in primitive reality, but the predictive success of physical laws does argue for there being such a model. And since physical laws express ordering principles, it is at least reasonable to assume that reality is a system of a not wholly random character. I will return to examine this idea in thermodynamic terms in a later section.

A second implication takes us towards the analogy I want to draw between beliefs in science and in theology.

A PARALLEL BETWEEN SCIENCE AND RELIGION

Fundamental to Christian thought is belief in the existence of God. Comparable postulation regarding a Transcendent Reality is intrinsic to any religious system, including those which do not mention God.[17] In any of its forms, this *existence postulate* is fairly widely conceded to be an expression of belief rather than of actual knowledge, though a few philosophers do argue that believers experience God's existence directly in a manner that parallels sensory (empirical) experience undergirding what I call knowledge in science.[18] The existence postulate brings God (or gods) into consideration *by hypothesis*. This can be understood in two basic ways in connection with the material world. On the one hand, it may be hypothesized that God is part of the reality of the world in which we find ourselves. A peculiarity of this view is that it leaves open the question as to the provenance of reality in relation to God. On the other hand, the existence postulate can be understood as an ontological *extremum postulate*, the ultimate existence being that of God. In terms of this extremum postulate, the world in which we find ourselves has the logical status of a secondary reality.

It has been borne in upon me over the years, as a working scientist, that very many of my colleagues—who knows how many?—would say that: *There is no scientific act of hypothesis parallel to the theological extremum postulate.* It is, as I have said, a main aim of this paper to argue against this negative view. And relevant to this main aim is a corollary of necessary realism brought out, ironically, in an especially sharp way by the Copenhagen disavowal of realism. According to the Copenhagen scheme, *it is impossible for any experimental observation to yield knowledge as to what character physical reality may have when it is not observed*. This is not merely a tautology; it points directly to the putative nature of a primitive reality. And this is precisely what was entailed when we said that the situation in physical science rests on the *fundamental assumption* that there exists a physical reality which is an independent and self-consistent system. The striking thing about this is that primitive reality thus enters into the picture in direct logical parallel to the *extremum postulate* in theology, for it is not simply that theory defines our beliefs about properties of a primitive reality. Because no physical experiment can prove the existence of a primitive reality, it is necessarily introduced as a fundamental assumption or belief.

I now turn to another branch of physics where paradoxes arise in connection with the fundamental concepts: thermodynamics.

A PARADOX IN COMMON INTERPRETATIONS OF THERMODYNAMICS

On my diagnosis, what goes wrong with the Copenhagen interpretation of quantum theory is the unwarranted *incorporation of the human presence in the statement of physical laws.*[19] Unfortunately, such a mistake is not confined to quantum physics. A quite different aspect of physics also treats human knowledge, or information, as a physical parameter. This is done through the idea, which seems to me to be widespread in cosmology and in modern chaos theory, that *entropy is the same as missing information.*

In classical thermodynamics we have the law of entropy increase, the second law of thermodynamics. On Kelvin's formulation, this expresses the law of degradation of energy: In closed systems, all energy initially present degenerates in time to the energy of heat. In standard modern formulations it is usual to express the law as the tendency for closed systems to approach maximum disorder. However, entropy is often understood as a mathematically defined quantity, the mathematical definition being the negative of what Shannon called *information* in his mathematical theory of human communication systems. The identification of information with entropy, on account of their theoretical models sharing formulae gives rise to a paradox.[20]

A supposed equivalence between entropy and missing information involves a confounding of properties of the observer with properties of observed physical systems. To be more precise, it involves the interpretation of the mathematical formalism in two categorically distinct domains of reference—the physical domain, in which the laws of thermodynamics are found to hold, and the domain of human cognition—and then ignoring the category distinction. If two or more categorically distinct entities have a common mathematical definition, this merely makes them analogous entities. It is a sheer error of logic to confound an analogic equivalence with an actual identity. How did this come about?

In physical terms, entropy is an additive quantity; when two systems interact the entropy of the one sums with that of the other. Boltzmann treated entropy in connection with the hypothesis that matter is of atomic composition. His fundamental concept is that of an *elementary disorder* which he puts into correspondence with the concept of *statistically independent events*. This introduced probabilities into physics. And, since entropy is an additive property of systems whereas the law of combination for statistically independent events is multiplicative, the only possible rapprochement is through the logarithmic function. The outcome is the Boltzmann relation:

$$S = k \ln W \qquad [4]$$

where S is the entropy, k is the Boltzmann constant—which may be set to unity—and W is a combinatorial term giving the number of possible events.

Shannon proved that when equation [4] is cast into probability form it provides a unique measure for the uniformity of any probability distribution. In that context the term S is quite reasonably understood as a measure of *uncertainty*, since an entirely uniform probability distribution indicates that any of the possible events is equally likely to occur. Naturally, uncertainty is reduced when relevant new information is available. The Boltzmann relation thus has interpretations in two *distinct* domains: thermodynamic entropy and human uncertainty. Conceptually, both entropy and uncertainty pertain to random situations and, hence, to the idea of disorder. But it is randomness or disorder in categorically distinct domains of reference.

If we now focus on the fact that both cases are defined by one and the same equation, it is then very easy to fall into the trap of saying that the introduction of further information reduces the uncertainty and, *thereby*, the entropy also. But this neglects a necessary distinction. In so doing, we make information into a *causal agent* that reduces the physical, thermodynamic entropy!

The supposed equivalence between entropy and missing information is, from the logical point of view, directly parallel with the Copenhagen treatment of human information or knowledge. In both cases, properties of the observer are confounded with properties of the physical systems under observation. In both cases, the ordering principles expressed in physical laws are founded in the human being in a way that leads to absurdity if imputed to an independent primitive reality. This absurdity can be avoided by attention to the ontological distinctions in the domains of application of thermodynamics and information theory.

An Oddity within Thermodynamics Itself

By bringing entropy into consideration, we have returned to the comment made earlier that the ordering principles expressed by physical laws presuppose primitive reality to be a system of a not wholly random character. This is an important consideration because the second thermodynamic law, the law of entropy increase, implies that disorder or randomness is also fundamental. It is therefore an oddity that the fundamental laws of physics should provide for order and disorder alike as fundamental, but without further comment as to why this should be so. I propose a new thermodynamic principle to resolve this oddity, which otherwise leaves thermodynamics looking incomplete.

To bring the oddity into focus, consider the second thermodynamic law in relation to the universe as a whole and, specifically, in terms of cosmic boundary conditions. A future boundary is provided if we take the view that the second law delivers the universe into its "heat death." But that outcome is an anticipation obtained without taking past boundary conditions into account. The latter ought to be taken into account not only on grounds of symmetry, but because it makes no sense to predict the outcome of processes for which earlier states go unspecified.

If we equate entropy with disorder, then the law of increase of entropy can be satisfied only if there is an initial *order* which goes into *decrease* under that law. This gives rise to the familiar question: Where did the order come from in the first place? Because that question goes unanswered by thermodynamics, it follows that the theory is incomplete in an *essential* way.

The concept of order as a physical quantity was proposed by Schrödinger. Starting with our equation [4], the Boltzmann relation, Schrödinger argued that if entropy S is a physical quantity, then its reciprocal should be a physical quantity also. Using the fact that the logarithm of a reciprocal equals a negative logarithm, he wrote

$$-S = k \ln W^{-1} \qquad [5]$$

and called -S negative entropy or *order*. Subsequently, this identification of order with negative entropy has become a standard view that underlies important developments in physics and other sciences. In particular, the idea that entropy is the same thing as missing information has its roots in equation [5].

In its most general form, however, the reciprocal of entropy is not given by $\ln W^{-1}$ but, instead, by $(\ln W)^{-1}$. If we then take the reasonable view that both infinite entropy and infinite order are meaningless as physical concepts, we may then write

$$IS = 1 \quad (I, S \neq \infty) \qquad [6]$$

where S is entropy and I is *order* defined as *reciprocal entropy*.

One is struck not so much by the simplicity of this equation as by its apparent triviality, for it seems to assert only what we have already stipulated, namely, that $I = S^{-1}$. However, the attached constraint that neither I nor S can be infinite means that neither term on the left-hand side of the equation can vanish. In physical terms this eliminates an important dilemma.

Physical theory cannot define everything, and so we have to deal with concepts that are understood as primitives. Indeed, it suffices for most purposes to treat *randomness* (that is, *disorder*) as a primitive, and to think of *order* as a departure from randomness. In that way, these two concepts are not conceived as polar opposites but, rather, as being intrinsically related notions. Through this relationship the notion of randomness seems more fundamental than the notion of order, the latter being viewed as a departure from the former. But this leads to the dilemma my proposal resolves. On the one hand, if order is a departure from randomness, we would then beg the question if we *begin* by stipulating the presence of order; so it would seem that we need randomness as a past boundary condition. On the other hand, the second thermodynamic law deals with disorder (that is, randomness) in a way that *presupposes* initial order.

I eliminate the dilemma through the circumstance that neither factor on the left-hand side of equation [6] can vanish. Thus, in connection with both past and future boundary conditions, equation [6] asserts that *it is impossible to have either a completely ordered or a completely disordered universe.*

This is a new thermodynamic principle which acts as a constraint that symmetrizes the past and future boundary conditions. In this way, it both escapes the dilemma and deals with the oddity which otherwise makes thermodynamics an incomplete theory. Whether the origin of the universe is understood in terms of the Big Bang or otherwise, the cosmological constraint expressed in equation [6] means that there is an initially broken symmetry. This implies a limitation on assumptions underlying the use of symmetry arguments applied to the earliest cosmic events.[21]

Physical theories and the laws of physics are purely human constructions. It might therefore seem at least reasonable, if not inevitable, that they should be formulated explicitly in terms of the human presence. I have tried to show that such attempts must fail if they do not simultaneously acknowledge the existence of an independent reality as the putative source for observed empirical regularities of the sort that physical laws and theories describe. The laws and theories of physics do not, on this view, describe reality. They advance what we believe about it in the light of current thought. The *cosmological constraint* postulated here is put forward in precisely that spirit.

SOME BRIEF REMARKS CONCERNING ORDER

Regarding the discussion in the section above, A PARADOX IN COMMON INTERPRETATIONS OF THERMODYNAMICS, it is essential to recognize that replacing negative entropy by reciprocal entropy is a great deal more than a mathematical formality. Apart from the implications already mentioned, the replacement disposes of the supposed equivalence between entropy and missing information as a physical principle widely appealed to in modern physics. That principle depends entirely on our identifying *physical reality* with empirical reality. The physical concept of *information*, which is specified in terms of human observers, and formally identified with *negative entropy*, is directly tied to human awareness in a way that would make it nonsense to project the information concept onto an independent reality. But I have already argued that physical concepts restricted solely to the domain of empirical reality cannot, without paradox, be incorporated as physical principles. In contrast, the cosmological constraint introduces the concept of *order* defined as *reciprocal entropy*; it is an essential feature that the cosmological constraint is explicitly postulated as a fundamental property of an independent reality. The cosmological constraint still allows uninhibited scope for applications of mathematical information theory in domains where that theory is appropriate, but the constraint leaves no room for the supposed equivalence of entropy and missing information to be used as a physical principle.

A second remark worth making is that acceptance of the Big Bang theory of the origin of the universe would imply that time began with that initial singularity. Now time is an ordering principle. The emergence of this ordering with the origin of the universe would satisfy the demand made by the cosmological constraint I am proposing for an initially broken symmetry only if time-reversal is excluded as real. This is a happy consequence because the order of time is also the order of causality. The *principle of causality* expresses an ordering relation: effects cannot precede their causes.

The causality principle conflicts with the idea of *actual* time-reversal, since the latter would mean that effects would then occur before their causes. However, if we can postulate that time-reversal, where it occurs within, say, mechanics, is purely a formal operation, permitted within the model that the laws of physics delineates but not intended to describe what is possible in reality, the *causality principle is preserved*. The formal operation of time-reversal then allows us to trace events retrodictively, in the direction from effects to their earlier causes, which in no way violates the principle of causality.

NOTES

1. [Reviser's Note (i): Preparing this paper for publication is something of a delicate task. Anonymous referees and the Editor had called for a number of significant revisions and clarifications to make the paper acceptable for inclusion in *Facets of Faith and Science. Volume 3*, but Professor Stuart died before he was able to attend to the task. Rather than see the paper dropped, the Editor asked me to see if I could revise it on Stuart's behalf. I have done what I can to clear up minor errors, to shorten the paper, and to bring Stuart's central argument into sharper focus. I have tried to put no words in his mouth I think he might disown, even where I have needed to add several fresh passages where referees wanted an expansion of some point or to see how it related to discussions elsewhere in the relevant literature. (I have no idea whether Stuart knew the additional literature I cite, so I have marked these items with an asterisk in the notes.) But I have occasionally taken the liberty of adding parenthetically what I hope is a useful explanatory or critical remark that I do not pretend Stuart would want to own.]

2. Of course, this does not mean that everything hypothesized is believed; one can merely entertain hypothetical ideas without believing in them. But at any rate, hypothesis does not immediately introduce knowledge.

3. N. Bohr, *Atomic Physics and Human Knowledge* (New York: John Wiley, 1958), 64; N. Bohr, *Essays, 1958-1962, on Atomic Physics and Human Knowledge* (New York: Interscience, 1963), 60.

4. Galileo seems to have got into trouble because in his writings he left little doubt that he took Copernicus's hypothesis literally.

5. *P. Duhem, *The Aim and Structure of Physical Theory* translated by P.P. Wiener (Princeton: Princeton University Press, 1906/1954).

6. It might be worth clarifying here, to save myself repeating a more cumbersome phrase, that when I talk of an "independent' reality, I mean a reality that does not owe its existence to human construction, though obviously our theories about such a reality would necessarily be human constructions. I do not mean to suggest by this phrase that reality is also independent of divine construction or sustaining. I do not think the argument based upon paradoxes, as I call them, in physical theory can help us discriminate which metaphysical theory to prefer from among all those which postulate a reality sufficiently independent of human thought to enable us to explain science's explanatory success.

[Reviser's note (ii): The most parsimonious metaphysics that can embed science's findings without derogation or distortion is, of course, materialism (see, for instance, *M. Bunge, *Scientific Materialism* [Dordrecht: D. Reidel Publishing Company, 1981]) or scientific realism (see, for instance, many of the contributions to *J. Leplin, ed. *Scientific Realism* [Berkeley: University of California Press, 1984] or *P. Churchland and C. Hooker, eds. *Images of Science* [Chicago: University of Chicago Press, 1985]). But materialism is not the only way to go. There is *Popper's three-worlds pluralism (K.R. Popper, *Objective Knowledge* [Oxford: Oxford University Press, 1972]; K.R. Popper and J. Eccles, *The Self and Its Brain* [Berlin: Springer International, 1977]), Lonergan's significantly renovated Thomism (*B.J.F. Lonergan, *Insight* [London: Longmans, Green and Company, 1958]), the phenomenological approach (*H. Jonas, *The Phenomenon of Life* [Chicago: University of Chicago Press, 1966]), and Foster's more recent idealism (*J. Foster, *The Case for Idealism* [London: Routledge and Kegan Paul, 1982]). My own preference is for a version of Whitehead's philosophy of organism (*A.N. Whitehead, *Science and the Modern World* [New York: Macmillan Publishers, 1925]; *A.N. Whitehead, *Process and Reality* [New York: Macmillan Publishers, 1929], corrected edition edited by D.R. Griffin and D. Shelburne [New York: Free Press, 1978]). For brief discussions of my preference see *T. Settle, "Van Rooijen and Mayr versus Popper: Is the Universe Causally Closed?" *British Journal for the Philosophy of Science* 40 (1989): 389-403; *T. Settle, "Fitness and Altruism: Traps for the Unwary, Biologist and Bystander Alike," *Philosophy and Biology* 8 (1993): 61-83; *T. Settle, "The Dressage Ring and the Ballroom: Loci of Double Agency," in *Facets of Faith and Science. Volume 4: Interpreting God's Action in the World*, edited by J.M. van der Meer (Lanham: The Pascal Centre for Advanced Studies in Faith and Science/University Press of America, 1996). It is important to notice that all these philosophical positions, like materialism, intentionally pay full respect to science.]

7. M. Born, *Einstein's Theory of Relativity* (New York: Dover Publications, 1962), 69-70.

8. W. Heisenberg, *The Physical Principles of the Quantum Theory* (New York: University of Chicago Press, 1930) Reprint (New York: Dover Publications, 1949), 3.

9. S.W. Hawking, *A Brief History of Time* (London: Bantam Books, 1988), 140.

10. *J.S. Bell, "On the Problem of Hidden Variables in Quantum Mechanics," *Reviews of Modern Physics* 38 (1966).

11. *D. Bohm, "A Suggested Interpretation of the Quantum Theory in Terms of 'Hidden' Variables. I and II," *Physical Review* 85 (1952).

12. There is no particular theological significance implied here. It is not uncommon in the recent literature discussing science's increasing explanatory success for a person to pour scorn on those denying scientific realism by saying that if scientific realism is false, science's success must be viewed (heaven forfend!) as a miracle. See, for instance, *H. Putnam, "What is Realism?" in *Scientific Realism* edited by J. Leplin (Berkeley: University of California Press, 1984).

13. [Reviser's Note (iii): The "we" in this paragraph is not universal. In recent philosophical literature on just this problem, there is a stout minority opinion against trying to explain science's explanatory success. See, for example, *B. Van Fraassen, *Images of Science* (Oxford: Oxford University Press, 1980); B. Van Fraassen, "To Save the Phenomena," in *Scientific Realism* edited by J. Leplin (Berkeley: University of California Press, 1984); B. Van Fraassen, "Empiricism in the Philosophy of Science," in *Images of Science* edited by P. Churchland and C. Hooker (Chicago: University of Chicago Press, 1985); *L. Laudan, "A Confutation of Convergent Realism," *Philosophy of Science* 48 (1981); *A. Fine, "The Natural Ontological Attitude," in *Scientific Realism* edited by J. Leplin (Berkeley: University of California Press, 1984). In brief, they argue two things. First, the inference to realism from empirical regularity is not deductive. This point is well taken in the literature: it is common for people to call the inference to realism "abductive" where they do not call it "inductive." Secondly, they argue that the assumption of an orderly reality would not *explain* the consistency of empirical experience in the sense of deductively implying it. That would require (at least) some extra, questionable, assumptions about the trustworthiness of human sensory apparatus.]

14. [Reviser's Note (iv): This is not a matter of strict implication—hence the quotation marks that I have added to the original text at this point. Even if the assumption of primitive reality were to explain predictive success—which it does not quite, though it furnishes elements for such an explanation—the inference from the thing explained to the truth of what does the explaining is not deductive.]

15. In C.I.J.M. Stuart, "Inconsistency of the Copenhagen Interpretation," *Foundations of Physics* 21, no. 5 (1991): 591-622, I argue independently of the considerations introduced here that the Copenhagen interpretation of quantum mechanics is *internally inconsistent*. It is especially significant that the inconsistency arises directly in terms of the Copenhagen disavowal of realism.

16. See, for example, any number of variants in *Leplin, *Scientific Realism*, and *Churchland and Hooker, *Images of Science*.

17. For an excellent discussion of the place of the Transcendent in the major world religions (though not necessarily in primitive religions) see *J. Hick, *An Interpretation of Religion* (New Haven: Yale University Press, 1989).

18. See, for example, Plantinga on the subject of properly basic beliefs (*A. Plantinga, "On Taking Belief in God as Basic," in *Religious Experience and Religious Belief* edited by J. Runzo and C.K. Ihara [Lanham: University Press of America, 1986]).

19. Other physicists have made a similar diagnosis and complaint. See, for example, *M. Bunge, "A Ghost-Free Axiomatization of Quantum Mechanics," in *Quantum Theory and Reality* edited by M. Bunge (New York: Springer-Verlag, 1967).

20. It would be inappropriate to go into this matter in a detailed way, for the issue is indeed complex and I have expressed my detailed views in C.I.J.M. Stuart, "Negative Entropy and Entropy as Missing Information," *Physics Essays* 4 (1991): 284-90.

21. For further discussion of this point see C.I.J.M. Stuart, "On the Completeness of Thermodynamics," *Physics Essays* 4 (1991): 142-43.

8

The Shroud of Turin: Resetting the Carbon-14 Clock

Thaddeus Trenn

HISTORICAL INTRODUCTION

Shroud of Turin

This case study deals with the complex and highly contentious linkage between the biblically established event of the resurrection of Jesus and the extant artifact known as the Shroud of Turin. Tradition maintained that this linen cloth was actually the burial shroud of Jesus Christ. The linen cloth, when extended, is 14′ 3″ long and 3′ 7″ wide, quite sufficient to cover both sides of the body from head to toe. The cloth's weave, material and pattern match products in common use in the Middle East during the first century. The linkage is reinforced by the presence of pollen samples from plant material characteristic of Palestine found in sediment layers from the Lake of Galilee belonging to the time of Jesus.

The most striking feature of the Shroud of Turin is the image as to both style and character. A photograph taken in 1898 by Secondo Pia revealed it to be the nature of a photographic negative, yet one with an exceptionally high degree of resolution permitting very detailed analysis. To this day, no one has been able to adequately ascertain just how this image was formed—an image that just will not go away.

This cloth was frequently subjected to candle smoke and other sources of carbon contamination during periodic public expositions over the centuries. A severe fire in 1532 in the Savoy family chapel, where it was housed at Turin, engulfed the folded cloth, causing molten silver from the casing to drip onto the linen, searing holes right through the cloth and leaving extensive damage quite visible today.

Edessa Connection

Normally, the Shroud is kept rolled up today, as is done with yardage goods. Early on, it had been folded, doubled in four, in the manner of a flag, with the countenance uppermost. Tradition and modern research traces the original folded cloth to Edessa, Turkey (modern Urfa, Turkey), where the apostle Jude Thaddeus was actively preaching and healing. After Jesus' death and resurrection, this Hebrew from Edessa was entrusted with the burial shroud, which he took as requested on a mission to heal King Abgar V of Edessa. Historically, the Abgar dynasty at the east Syrian kingdom of Osrhoene is considered a landmark, being at the head of a long line of converted kings bringing Christianity to all their subjects. The cloth itself was rediscovered about AD 525 sealed for protection in the city walls of Edessa.

The Mandylion, as it was later named, was brought to Constantine VII in AD 944. The French Crusader Knights stole it during the sack of Constantinople in 1204, and brought it home where it turned up in Lirey as a spoil of war. Eventually it was transported to the ruling House of Savoy, based in Turin, where it was housed initially in the family chapel. From 1694 it was housed in a new chapel built specifically for this purpose. The Mandylion was long known to be more than merely the evident facial image visible atop the packet. Examination of the full image reveals the front and dorsal aspects of a crucified man, brutally lashed and crowned with thorns. The location of the wounds, including the unexpected position of nail holes—in the *wrists*, not palms—as well as the open gash at mid-body conform with astonishing precision to biblical reports. The mummy wrap tradition, long used to depict Christ's burial, had already begun to change by 1025 in Constantinople to correspond with a full-length draped cloth. This reflects full awareness of the meaning and physical characteristics of the Shroud.[1]

A Problem for Authenticity

Ancient tradition venerated the cloth's image as *Acheiropoietos*, "made without hands." During the fourteenth century a controversy erupted suggesting that this only *looked* like a burial garment but was in fact merely a work of art *cunningly* produced. This infamous D'Arcis affair has been diffused historically as being a manifestation of acrimonious interpapal rivalry ongoing at the time.[2] Modern science has confirmed that it is manifestly not a work of art of any type. The image is confined to the surface and appears as yellow discolorations of the cellulose—there is no penetration of the fibres as would be expected from any artistic medium.[3] Yet microscopic photographs verify the presence of bloodstains which have seeped into the crevices of the fibres in the image areas. Most baffling, the image has a very definite three-dimensional character.

But if it is a burial garment, as scientific evidence now seems to guarantee, how did the image form on the cloth? One might wish to conclude that this was somehow an effect of the resurrection event, but two problems intervene at this juncture. First, there is no known process readily available to make the connection scientifically. Second, and quite serious from the point of view of authenticity, is the recent radiocarbon dating purportedly breaking the link with Jesus altogether. Unless the fibres forming the cloth existed at least 1960 years before the present, this clearly could not have been the burial garment of Jesus Christ. Carbon dating in 1988 claims an age less than 700 years! The claim was immediately accepted, albeit uncritically, by the public and most scientists alike. Perhaps the cloth was the burial garment of a fourteenth-century Crusader Knight captured by the Saracens and crucified exactly in the manner of Jesus in deliberate mockery of the latter's death.

RESETTING THE CARBON-14 CLOCK

Carbon Dating

Assuming the carbon reservoir to be in a state of equilibrium, carbon taken up by metabolic processes will remain in the same ratio until the time of death. At that point, the decay of carbon-14, radiocarbon, makes the ratio it has with stable carbon-12 decrease slowly with time, inexorably like clockwork.[4] The date of death can then be deduced, accordingly, in a comparative fashion, with some considerable degree of accuracy. Carbon-14 is produced mainly in the upper atmosphere. Neutrons of moderate energy, released by cosmic radiation, are captured by nitrogen, the most abundant gas. Nitrogen-14 loses a proton in the exchange, effectively reducing its atomic number one unit. The resulting carbon-14 leaves its tracks in the form of electron emission, as it undergoes weak beta decay, transforming itself back into nitrogen-14, taking about 5,700 years for half of it to change.

Now, to determine the carbon ratios, it is far more efficient to count carbon atoms directly instead of waiting for radiocarbon to decay. Greater resolution is achieved by initially accelerating the isotope ions to *high* energy. The advantage of using this so-called *accelerator* mass spectrometer (AMS), besides the higher resolution that is achieved, is that very small specimens are required for destructive testing in comparison with what is needed for radioactive techniques. A *sine qua non* for accurate carbon dating is that no secondary radiocarbon be present to skew the ratio, which is particularly sensitive to this carbon-14 component, normally present at less than one part in a billion! With some materials, it is often difficult to distinguish carbon contaminants, which include carbon-containing compounds that are not indigenous to the original sample matrix, but which have become physically or chemically fixed as part of the specimen.

Carbon Contaminants

Researchers dealing with the Shroud of Turin were keenly aware of potential sources of carbon contamination and tried their utmost to eliminate these anticipated contaminants during pretreatment cleaning. It is well known that the cloth in question came into contact with smoke periodically over the centuries.[5] Such surface contamination might well have become etched, as it were, into the fibres during the serious fire of 1532. If not eliminated or calibrated out, excess radiocarbon as contaminant would have the effect of shortening the calculated age. According to one estimate, it would take only eighteen percent excess of carbon-14, whether as a contaminant or produced *in situ*, to shift the date from 2,000 years to 650 years before the present.[6]

The AMS technique is particularly troublesome with regard to accuracy, as is well known, normally requiring the control of more traditional procedures except for estimates. Anomalous results obtained have been duly noted, and traced in many cases to the pretreatment of samples tested. So clearly there were serious potential costs accompanying the benefits expected when only three AMS-oriented laboratories were selected to test the Shroud of Turin: namely, Zurich, Arizona and Oxford.

These three laboratories each treated their shroud sample as a whole piece of cloth, rather than shredding or unravelling the fibres, because otherwise pretreatment cleaning would be more difficult. Such treatment was crucial, since the shroud had been exposed to a wide range of potential sources of contamination; dirt, smoke and other contaminants to an unknown degree. The laboratories each subdivided their own sample and then used several mechanical and chemical cleaning procedures. Each laboratory was provided with a sample of three control specimens of known origin and date, none of which was linked with such an indeterminate background of contamination.

Radiocarbon Dating Moot

Their findings were published in *Nature*.[7] "The results provide conclusive evidence that the linen of the Shroud of Turin is medieval." [sic] "The age of the shroud is obtained as AD 1260-1390, with at least 95% confidence." As it stands, the evidence appears unassailable. Yet the interpretation of the results, upon closer scrutiny, is clearly subject to doubt.

Each laboratory reported "no significant differences between the results obtained with the different cleaning procedures that each used," and, taken together, each result was in the same range. But this implies that the elaborate chemical procedures applied by each of the laboratories were no more effective in the removal of contaminants than were simple physical procedures. The Zurich subsample, designated Z1.1u, "received no further treatment" beyond the precleaning ultrasonic bath. Yet, without the benefit of any chemical treatment this subsample actually yielded the *older* datum of the several runs! Arizona utilized various *chemical* pretreatment procedures exclusively.

It strains scientific credulity to suppose that simple *physical* procedures and complex *chemical* procedures, each applied independently, could just be interchangeable in their effectiveness. At the very least, this would render the elaborate chemical pretreatment procedures superfluous. Logic dictates that the procedures actually utilized in this case were simply not differentially effective in removing carbon contaminants. Either there were no such contaminants in the first instance or else the results obtained require revision. Yet the researchers involved acknowledged the checkered history of this linen cloth, exposed as it was to smoky environments over time. The intense fire of 1532 alone dictates the need for pretreatment cleaning. The procedures selected, unfortunately, cannot be deemed successful.

FURTHER RESETTING THE CARBON-14 CLOCK

In Situ *Production of Radiocarbon*

Besides contamination there is another way by which excess radiocarbon might well have become fixed in the fibres, immune from pretreatment cleaning. If the atmosphere did not attenuate the range of cosmic radiation, the conversion of nitrogen-14 to carbon-14 could easily occur on earth, *in situ*, as it were, affecting samples both living and deceased. The same result would occur in the presence of a neutron flux from some other origin. Nitrogen-14 fixed in the fibres, in the presence of a neutron flux, could undergo transformation to carbon-14 just as in the upper atmosphere. The AMS team at Oxford actually anticipated this theoretical possibility.[8] Higher yields of radiocarbon would be expected from this standard *nitrogen* process as compared with the equally plausible neutron activation of carbon-13. Now, it is doubtful that the Oxford team truly believed such a neutron flux actually constituted part of the historical environment of the Shroud of Turin. The point to be made here is simply that *if* one *were* to have occurred, then the *in situ* effect upon nitrogen-14 fixed in the fibres would require attention.

The source of this idea can be traced to the suggestion by several individuals that a burst of radiation accompanying the resurrection event might be one way to account for the formation of the image on the linen cloth. One variation of this proposal includes a neutron flux to produce excess radiocarbon. While this may yield a result helpful for elucidating anomalies regarding the Shroud of Turin, there is a fundamental reason for excluding any such neutron event. If it were introduced as a singleton event, unique in all history, then it would be considered untestable in principle and would elude scientific scrutiny altogether. This is why it was considered so important that pretreatment procedures should have been adequate to eliminate any possible excess radiocarbon produced *in situ*. Although radiocarbon is radiocarbon however it is produced, the precleaning techniques employed were considered quite adequate to *selectively* remove any hypothetical excess carbon-14 in virtue of the chemical location of the fixed nitrogen which would have been involved. Since similar results were obtained for both chemical and physical techniques, it was presumed that no such hypothetical excess carbon-14 was even present in the first place.[9]

Although the carbon dating investigators presumed that any excess carbon-14 would be selectively removed by pretreatment cleaning, there is good reason to doubt that such secondary radiocarbon would be separable. As contamination from smoky environments, excess radiocarbon *might* be separable, but not easily if it were etched into the fibrils by excessive heat. The lack of any difference obtained, using physical methods or chemical pretreatment cleaning, suggests that any such extra carbon-14 was merely added to the total carbon-14 content *a fortiori* foreshortening the date obtained.

Radiocarbon could also be induced from neutron irradiation of *in situ* nitrogen-14. The Shroud investigators considered that such nitrogen-14 would be in a chemically different environment. If so, any carbon-14 arising from this nitrogen-14 would be in a chemically different environment. Such carbon-14 was presumed to be susceptible to selective removal by the chemical cleaning techniques. Yet they must still account for the lack of any advantage for such *chemical* procedures over the simple physical ultrasonic bath as discussed above. But the chemical environment may *not* be different! Jitse van der Meer has suggested that any nitrogen-14 would be found only in the proteins, namely the same chemical location as for photosynthetic carbon-14 of original endowment: "So the two carbon-14's are in the same chemical environment within the protein molecule."[10] If true, this would undermine their presumption of selective removability of any excess carbon-14 and would strengthen the central argument of this paper, namely, that excess carbon-14 was inevitably co-present throughout the carbon dating investigations.

Both alternatives lead to the same basic conclusion, namely that pretreatment cleaning was ineffective at removing induced radiocarbon, if indeed any had been induced. Whereas excess radiocarbon (as contaminants or from neutron flux) could not be easily *removed* selectively vis-à-vis radiocarbon of original endowment, it is of course, entirely possible to test the overall density distribution of the *total* carbon-14 content. Assuming the distribution of photosynthetic radiocarbon to be uniform throughout the cloth, any lack of uniformity overall would strongly indicate the presence of excess carbon-14. Induced excess radiocarbon would be commingled and superimposed upon this background of photosynthetic radiocarbon. But unless the neutron-induced radiocarbon was quite random and uniform in distribution, the *total* count, when scanned across the width, would not be uniform. It is this possible discrimination regarding the total carbon-14 content, by means of counting differentials, that is our concern. For this would provide a *direct* indication of *non*uniformity in the overall radiocarbon distribution. The implications of this startling result should be obvious even to the skeptic.

Effect of Additional Radiocarbon

While the uniformity of results obtained by the three laboratories was interpreted as evidence for lack of any residual excess radiocarbon, another conclusion is possible. If some excess radiocarbon were fixed in the fibres in such a manner as to be indistinguishable from the original endowment—a well known source of other dating errors in the past—then the absence of detectable difference might be turned around to suggest that the pretreatment techniques used were simply inadequate. Perhaps some secondary radiocarbon merely eluded the procedures. The argument against the presence of secondary radiocarbon formed by "neutron capture by nitrogen in the cloth" was based upon the "equivalent results" obtained by the three dating laboratories using "different types of chemical pretreatments."[11] But it is now clear that these elaborate chemical procedures could have been replaced altogether with a simple physical process (ultrasonic bath) vis-à-vis their alleged effectiveness at removal of carbon contaminants. So it is questionable whether these same chemical procedures could have been any more effective at discriminating the separation of secondary radiocarbon from the original endowment, notwithstanding the claim that "the original nitrogen is in a chemically quite different environment." Where detectable difference is absent, prudence would suggest that it would be wise to begin by questioning the effectiveness of the procedures selected.

The problem for the radiocarbon dating results is evident. Any significant excess radiocarbon, normally present in the isotopic mix known as carbon at less than one part in a billion, would quite radically skew the ratio between carbon-14 and stable carbon-12. Accordingly, as the relative amount of radiocarbon increases, the age of the specimen will appear younger than it ought to be if given the original endowment. The presence of radiocarbon produced *in situ* would scramble the carbon dating clock more seriously than carbon contaminants.

Implications of Excess Radiocarbon

Normally, only a negligible amount of radiocarbon would be produced in any given specimen due to cosmic radiation reaching the earth. The enhanced amount envisioned here as the result of a neutron flux has wider implications beyond modifying the carbon ratio, perforce foreshortening the measured age of the cloth. Such a neutron flux would undoubtedly be an accompaniment of some physical process as yet to be determined. Parsimoniously, such a process would also involve the cause of the image formed on the cloth. It ought as well to deal directly with the resurrection event itself, to the extent that this is at least in part also a manifestation of some physical process.

The problem here is that while it might be possible to explain the dating anomaly through the introduction of excess radiocarbon caused by a neutron flux accompanying a resurrection event that also produced an image on the cloth, the very nature of the event, however plausible it may be scientifically, tends to elude scientific methodology. Science is indeed at its best dealing with reproducible events, although it is making headway on possibly the most glaring exception, with ever increasing theoretical understanding of events associated with the Big Bang.

Irreproducible events do occur. Science cannot assert that a given process or event did *not* occur simply in virtue of poverty in probability. In order even to begin to approach the mystery of the Shroud of Turin, I believe that we must adequately take into consideration the complex constellation of events associated with it—however rare or unique—events which, while distinct, are nevertheless collectively as unique as the Shroud of Turin itself. For if Jesus Christ was indeed buried in this linen cloth and then resurrected while wearing it, this is surely of great consequence.

RESURRECTION MODEL

Dematerialization

I should like to introduce weak *dematerialization* as perhaps a key feature of the resurrection event associated with the Shroud of Turin. This is essentially different from the case of Lazarus.[12] Let us take the example of boiling water whereby molecular bonds are overcome, freeing the gaseous products in a vapor we call steam. By analogy, at the nuclear level, the strong force bonding the main units—neutrons and protons—is expressed in terms of exchanging a particle known as pi-meson, or pion. If this pion *glue* were overcome by energy supplied to the system, we could speak of a sort of phase change, as when water turns to steam. One striking feature of pion bonding is the missing mass. The slight mass differential between paired and unpaired nucleons, neutrons as well as protons, is termed *mass defect* in the language of mass. Energy-wise it is known as *binding energy*.[13] So, energy would be required to violate those unions—energy pumped into the system, as it were, to replace the *binding energy*.

Each nucleus is *missing* about one percent of its mass. For an object weighing, say, eighty kilograms, the missing energy would be the equivalent of nearly thirty-six grams of matter converted by $E=mc^2$. Put graphically, to resurrect a body this way would require the energy nominally equivalent to about twenty-nine atomic bombs. Here it is crucial to realize that we are dealing *not* with energy released but with energy to be *replaced*—the extra energy to break those pion bonds. By analogy, an exothermal chemical reaction liberates heat, whereas an endothermal one absorbs heat. Clearly then, a resurrection event accompanied by *weak* dematerialization would not devastate the tomb. Robinson has suggested "transmaterialization" as an alternative in order to capture the traditional glorified-body aspect without depending upon any physical explanation.[14]

Singleton Event Complex

Once normal pion bonding were overcome, the nucleons of matter would have been set free. These would subsequently behave according to normal physical laws. Almost instantaneously the nuclei would begin to lose their orbital electrons. Electron orbit-jumping is normally accompanied by soft x-rays characteristic of the shift. These would be sent speedily from every element of the body surface. The electrons, now unbound, would tend to fly to ground—literally and electrically. The dry cloth would act in the first instance as earth potential. As with the x-rays, these electrons would also emanate from every atom of the body, streaming away as if collimated, which is normal to the body surface in the manner of coronal discharge. Aside from the obvious advantage to image resolution if each atom contributed to the process, recent scientific research supports the position that soft x-rays and coronal discharge could yield the sort of image effect produced by the Shroud of Turin.[15]

The unbound neutrons would yield a dense flux throughout the length and width of the cloth impacting upon nitrogen-14 atoms fixed in the fibres with energy sufficiently moderate to convert many nuclei into radiocarbon. These neutrons would no longer be "forbidden" to transform themselves into protons, so the proton plasma would continually grow and disperse itself throughout the atmosphere within the tomb. The burial cave hewn out of rock is suddenly filled with hydrogen mixed with oxygen. The slightest additional spark would have set off a most violent explosion within the confined space, generating water, and forcing aside by concussion any object in the way. The linen cloth would have been violently blown aside, and the stone would have been dislodged from the entrance. Anyone subsequently entering the empty tomb would have found the linen cloth tossed aside, but bearing signs of an image. Those witnesses, though, could not have known about the excess radiocarbon fixed in the rock—kept secret until our day.

Nonuniform Density of Radiocarbon

The original endowment of radiocarbon gained through metabolic processes would be present uniformly throughout the entire cloth. Any secondary radiocarbon produced *in situ* would tend to vary in density as a function of the neutron flux distribution. Various attempts have been made to account for the cloth's image in virtue of a burst of radiation. Some have included a neutron flux as an accompaniment of this burst, perhaps in the manner of a neutron bomb yielding high velocity neutrons.[16] The suggestion here of weak dematerialization, achieved through overcoming pion bonding of the nucleons of matter, is very different. The neutrons would simply be freed to stream through the cloth with only moderate energy. Accordingly, we would expect a variation in the density distribution, highest in the center of the cloth along its entire length, namely, where the body lay. It would be simple to test this prediction of an axial maximum.

Recall that radiocarbon transforms itself back into nitrogen-14 through a process of weak beta decay, taking about 5,700 years for half to change. Less than twenty-five percent of this secondary radiocarbon, produced *in situ* by this alleged neutron flux, would have undergone weak beta decay to date. Beta rays are electrons from the nucleus, and are exceptionally weak in the case of radiocarbon. But if this nuclear electron emission were initially accelerated by means of a positively charged fine mesh grid especially placed against the spread-out Shroud for testing purposes, the impact distribution could be coherently registered by any number of available detection processes. Alternatively, a matrix of photographic plates with lead sheets sandwiched in between could be enfolded within the Shroud in total darkness for a period of time to clearly differentiate the frontal and dorsal aspects of the variation in density distribution, obtaining a visual record for detailed optical scanning. So, although the Shroud of Turin may elude the grasp of scientific methodology in virtue of the irreproducibility of the event complex that led to the evidence to hand, it may nevertheless yet have something to teach science.

Concluding Postscript

Uncritical Rejection

I have long been intrigued by the strong tendency to dismiss the Shroud of Turin as a forgery or fake of some type. Typical reactions are neither neutral nor objective as to the question of its authenticity, but tend to be negative and sceptical, not infrequently virulently so. What is it about this supposedly simple linen cloth that evokes such hostility? Could it perhaps be a vaguely felt fear of the unknown? A common sense intuition that there is far more to this particular linen cloth than meets the eye? An escape to the dungeon of unbelief to avoid the blinding light? That the Shroud of Turin is an extant artifact is an unassailable fact that has long attracted world attention. Why, though, is the public so inclined to be more comfortable with claims that it is not what it purports to be? True, many people were no doubt disappointed in 1988 to learn that radiocarbon dating had placed it in the fourteenth century. Scientifically, though, the conclusions went far beyond what the evidence would bear. The presence of radiocarbon was actually found to be in quantity inconsistent with the date of 2,000 years before present and consistent with the fourteenth century *provided* no secondary radiocarbon had been introduced during the intervening years. This latter proviso was not even specifically stated, much less considered, although given the tradition associated with this linen cloth, it is historically questionable to dismiss it, even though this move conforms with sound scientific methodology. Beyond the exaggerated claims of science came the public media with unsupported implications that "fake," "forgery" and "fraud" are somehow appropriate words to describe this artifact. It is just this "jumping to conclusions" and uncritical rejection that seems to me the litmus test of authenticity. For *science* has already confirmed that this cannot be a work of art of any sort, and *science* stands perplexed, without as yet an explanation for how the three-dimensional image was formed. Some fake!

Blinded by a Parable

Throughout his earthly life, Jesus often spoke in a parable which in Aramaic may be translated as "riddle." How apropos that he should have left such a magnificent conundrum for posterity—an enigma tantalizing for science, but just beyond its grasp—a playful yet powerful parable to the sceptics of our day who examine closely with eyes enhanced by science, though still fail to see. For if Jesus Christ was buried in that linen cloth, and was resurrected while wearing this underdeveloped film, then these events *must* perforce be taken into consideration when scrutinizing the Shroud of Turin. To construe it as merely another piece of cloth is to fall victim to this ultimate parable. Yet science, in virtue of its mentality and method, can forever see only a cloth, not the Shroud of Turin:—*a fortiori*—neither can it adjudicate its authenticity.

Yet irreproducibility aside, a considerable amount of scientific evidence and data has been accumulated over the years concerning the Shroud of Turin. And the linkage between image formation, excess radiocarbon and a resurrection event involving weak dematerialization accompanied by a neutron flux does lead to a testable prediction as regards the expected axial maximum in the density distribution of secondary radiocarbon. Still, such an alternative account, however plausible scientifically, takes us into the realm of belief and unbelief at a very deep level, exposing hidden motivations and attitudes which in turn condition the operation of presuppositions and theory formation including the acceptance or dismissal of "evidence." In the absence of a dispositional attitude of belief, there will be no way to "prove" that this might be the case. The mocking assertion by one of the principal scientists involved with the radiocarbon dating is but one indicator of such attitudinal unbelief: "Someone just got a bit of linen, faked it up and flogged it."[17]

Authenticity Reconsidered

The immediate issue of authenticity concerns whether this is indeed a burial cloth of the first century. Of course, if it is authentic, it would not "prove" the Resurrection. Belief in the Resurrection must not depend upon this burial garment or anything else. Nevertheless, unconditional belief in the Resurrection event, *whatever* may actually have taken place, may well dispose the heart and mind to at least accept the possibility of alternative processes and hypotheses which may well be ancillary towards an understanding of the Turin Shroud. Without such an attitude of openness, in catch-22 fashion, we may exclude even from consideration the very sort of event-complex that could substantiate the authenticity of the Shroud as a first century garment.

The answer to the question of the Turin Shroud's authenticity is written on the heart, and quite eludes the scope of science to definitely affirm or refute. Scientism is our modern ideology. Much like Thomas the doubter, we refuse to believe what eludes sense and reason. Following Adam, we have pridefully succumbed to the lure of our modern tree of knowledge only to find its fruit bitter to the taste. How pretentious to expect that scientific evidence could provide the easy way—a *firm* foundation for belief. The Shroud of Turin is a mystery. Merely dating its fibres at 2,000 years will not guarantee its authenticity. Nor will a sound theory of image formation resolve the conundrum. Even the results of future tests on the density distribution of secondary radiocarbon will leave sceptics in its wake. But neither can the recent radiocarbon dating results be accepted as a verdict of dismissal. The resurrection event is probably the key to this enigmatic bequest. To interpose our science as an adjudicator of belief is pride of the rankest sort. The essential act of faith in Jesus Christ must not be mediated or made contingent upon scientific findings or anything else for that matter.

Without such unconditional belief, the Shroud of Turin remains quite inaccessible. But believing in the Shroud for its own sake could turn it into just another golden calf.[18] Faith in Christ must not be dependent

on *any* sort of idol, sacred or profane. Whatever function it might
have, the Shroud must not become the object of our faith. Nor may we
allow it to be construed as a condition or test for faith. It cannot be a
crutch or substitute for the essential act of faith. Yet, if rightly
understood, it may for some who already believe unconditionally, even
reinforce and further validate the earthly claims of Jesus Christ, no less
than his many miracles, performed in the presence of faith.

NOTES

1. K.E. Stevenson and G.R. Habermas, *The Shroud and the Controversy*
(Nashville: Thomas Nelson, 1990), chap. 4 and passim; I. Wilson, *Holy Faces,
Secret Places* (New York: Doubleday, 1991), chaps. 10-14 and passim;
F.T. Zugibe, *The Cross and the Shroud* (New York: Paragon Press, 1988),
chap. 12 and passim.

2. Wilson, chap. 2; Stevenson and Habermas, 155.

3. Stevenson and Habermas, 31, 121, passim.

4. R.E. Taylor and R.A. Müller, "Radiocarbon Dating," in *McGraw-Hill
Encyclopedia of Science and Technology*, vol. 15, 6th ed. 1987, 121-28.

5. P.E. Damon, D.J. Donahue, B.H. Gore, A.L. Hatheway, A.J.T. Jull,
T.W. Kinick, P.J. Sercel, L.J. Toolin, C.R. Bronk, E.T. Hall,
R.E.M. Hedges, R. Housley, I.A. Law, C. Perry, G. Bonani, S. Trumbore,
W. Woelfli, J.C. Ambers, S.G.E. Bowman, M.N. Leese, and M.S. Tite.
"Radiocarbon Dating of the Shroud of Turin," *Nature* 337 (1989): 611-15,
612; Wilson, 176ff.

6. B. Kelly, "Turin Shroud," *New Scientist* 119 (1988): 94; dates are
indicated backwards from present.

7. Damon, et al.

8. R.E.M. Hedges, "Reply," *Nature* 337 (1989): 594.

9. "The three laboratories used different types of chemical
pretreatments...yet obtained equivalent results. This shows that any C^{14} formed
by neutron irradiation behaves chemically in the same way as the original C^{14}.
This is inherently unlikely because the original nitrogen is in a chemically quite
different environment." (Hedges, 594).

10. J.M. van der Meer, personal communication.

11. Hedges, 594.

12. J.A.T. Robinson, "The Shroud and the New Testament," in *Face to
Face with the Turin Shroud*, edited by P. Jennings (Oxford: Mowbray, 1978),
79.

13. K.W. Ford, *The World of Elementary Particles* (New York: Blaisdell, 1965), 174 and passim. One way to envision the energy influx is via very high energy gamma rays focused on the nuclei. A far simpler way would involve premature and spontaneous pion decay, whereby "leashed" *virtual* pions escape as *real* pions which then automatically decay in a fraction of a second. Compare Ford, 195, 207-9, passim. Weak dematerialization is like "Lego" blocks which are being disassembled but do not disappear.

14. Robinson, 79.

15. Stevenson and Habermas, 40-41 and passim; Zugibe, 178. As for hypothetical carbon-14 fixed in the fibres, compare quote at note 9. One referee noted that the suggested process might have also left some traces in the rock of the tomb, and that in principle one could make relevant measurements at the Holy Sepulchre in Jerusalem.

16. T.J. Phillips, "Shroud Irradiated with Neutrons?" *Nature* 337 (1989): 594; Stevenson and Habermas, 167, 198-99, 204; Zugibe, 173-78.

17. Wilson, 5, quoting Professor Edward Hall.

18. Stevenson and Habermas, 146-47, 169.

9

Rationalism, Voluntarism and Seventeenth-Century Science

Edward B. Davis

INTRODUCING THE PROBLEM: ON INTERPRETATIONS OF EARLY MODERN SCIENCE

"The relation of science to religion in the seventeenth century," writes Richard S. Westfall, "is the central problem in the history of modern Western thought."[1] With this bald assertion I wholeheartedly agree, yet I must reject the terms in which it is couched. It was religion, not science, which held the dominant cultural position in seventeenth-century Europe, so the implied question behind Westfall's claim ought to be turned around. What was the relation of religion to science in the seventeenth century?

Sociological approaches to this question have been the subject of a veritable mountain of research, much of it growing out of Robert Merton's work more than fifty years ago on the influence of Puritanism on scientific activity in England. The sheer amount of attention paid to this particular aspect of a book that is much broader in scope has surprised Merton himself. "Had educated and articulate Puritans of the seventeenth century been social scientists," Merton reflects in 1970, "they would have found this focus of interest passing strange," for they assumed almost without question that science was an equal partner with theology in their search for truth.[2] Yet what we have grown accustomed to calling the "Merton thesis" did not at first seem very persuasive. By Merton's own admission, many scholars saw it as "an improbable, not to say, absurd relation between religion and science." John William Draper and Andrew Dickson White, he notes, had convinced a generation of scholars that conflict between religion and science was a "logical and historical necessity," so that "a state of war between the two was constrained to be continuous and inevitable."[3]

Merton's diagnosis was largely correct. It is only in the past twenty to thirty years that scholars in significant numbers have taken explicit exception to the warfare school.[4] But if Merton was bold to challenge the dominant view of the relation between Christianity and science, he was not alone. Though Merton did not know it, at the same time that he was engaged in his research on the influence of the Puritan *"ethos"* on the *reception* of science, the late British philosopher Michael Beresford Foster (1903-1959) was writing about the influence of Christian *theology*, especially the voluntarist variety, on the *content* of early modern natural philosophy.[5]

Foster makes his long, convoluted, often confusing argument in a series of three articles published in the journal *Mind* in the mid-1930s, early in his career as a student (read, "tutor") at Christ College, Oxford.[6] He focuses on the doctrine of creation as the vehicle through which theology impinges on natural philosophy, a particularly promising approach because most natural philosophers of the early modern period believed without question that the world and the human mind had been created by the omnipotent God of the Judeo-Christian tradition. The classic triangle of relations among God, nature and the human mind determined the way in which they viewed scientific knowledge (Figure 9.1): both the manner in which and the degree to which the universe could be understood depended on how God had acted in creating it and how God continued to act in sustaining it.

Over the centuries, Christian thinkers, though reaching a consensus on the reality and goodness of the creation, have differed widely on the precise nature of the created order. Nevertheless, it is possible to say in general that the history of the Christian doctrine of creation often involves a dialogue between an emphasis on God's unconstrained will, which utterly transcends the bounds of human comprehension, and an emphasis on God's orderly intellect, which serves as the model for the human mind. This dialogue rose to prominence in the late Middle Ages, when theologians developed the distinction between God's absolute power (*potentia absoluta*) and God's ordained power (*potentia ordinata*). In the words of William Courtenay, this means "that according to absolute power God, inasmuch as he is omnipotent, has the *ability* to do many things that he does not *will* to do, has never done

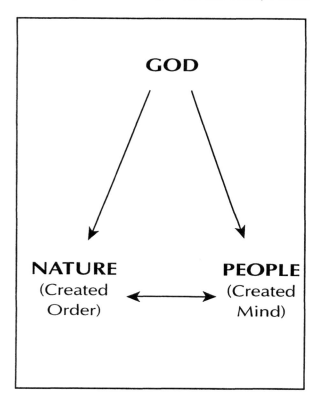

Figure 9.1: Relations among God, nature and the human mind.

nor ever will do." Medieval theologians used this distinction to affirm divine freedom in creation, and to underscore the contingent, covenantal character of all created things.[7] Early modern Christian thinkers worked within the linguistic boundaries of a similar dialectic. Typically they would acknowledge that God has both will and reason, but then stress one more than the other. Those emphasizing the divine will are often called "voluntarists," and those emphasizing the divine reason are often called "rationalists."[8]

Foster's work employs this dialectic. "The method of natural science," he argues, "depends upon the presuppositions which are held about nature, and the presuppositions about nature [depend] in turn upon the doctrine of God." Foster identifies two basic attitudes toward

God which differ substantially in their implications for scientific methodology. Rationalist theology "is the doctrine that the activity of God is an activity of reason." Since "God is nothing but reason, there is...nothing mysterious or inscrutable in his nature." Such a theology, Foster says, "involves both a rationalist philosophy of nature and a rationalist theory of knowledge of nature." As a product of divine reason, the world must embody the ideas of that reason: and our own reason, "in disclosing to us God's ideas, will at the same time reveal to us the essential nature of the created world." A voluntarist theology, on the other hand, "attributes to God an activity of will not wholly determined by reason." Thus the products of God's creative activity are not necessary, but contingent. Since our minds cannot have demonstrative a priori knowledge of a contingent reality, the created world can be known only empirically.[9]

For Foster, the connection between Christian theology and the presuppositions of modern science is itself logically necessary, not historically contingent.[10] It seems fair to say that his articles go beyond historical revisionism to the point of engaging in apologetics: by showing the fundamental significance of Christian theology for modern science, he hopes to use the prestige of the latter to benefit the former. This approach has provoked an important critique by the German theologian Rolf Gruner, who does not share the high view of science held by Foster as well as by many opponents of Christianity. Gruner, hoping to put a stop to the apologetic uses of revisionism, takes issue with all those who see Christianity as the source of modern science. It would be more promising, he says, for theologians in the future "to further the prestige of their religion by maintaining that it demands man's respect for his so-called environment rather than its manipulation and control."[11]

This is a point well worth making—science and technology are no longer viewed uncritically as unmixed blessings in many circles[12]—but it fails to touch the substance of Foster's argument, and it can only preach to, not argue with, those who do not share Gruner's negative opinion of modern science. Gruner does raise several other objections to the Christianity-and-science thesis, including some that might be called historical in the broadest sense.[13] However, most of his objections are theological, and none of them confronts the actual historical situation in the seventeenth-century, when what we now recognize as modern science came slowly into existence. Ultimately Gruner realizes that the revisionist case "is really the case of Bacon refurbished for the twentieth century,"[14] in that revisionists are simply repeating Bacon's attempt to show that the new science was closely tied to Christian assumptions. But this is to concede the very historical point

I want to make: the revisionist case remains attractive to many precisely because it does ring true for certain aspects of seventeenth-century natural philosophy.[15]

Perhaps the most serious objection to Foster's work, however, is that his articles are profoundly unhistorical, dwelling in an abstract space somewhere above the plane of real thinkers constructing real natural philosophies. Although he argued vigorously what a consistently pursued theology of creation *ought* to have entailed for early modern natural philosophy, he does little to show what has actually been the case. Yet he focuses our attention on two of the right questions: How in fact did early modern thinkers construe the relation between God and the creation? What did they think this meant for human efforts to understand the created order? Picking up on Foster's essential insights, a growing number of scholars have begun to fashion a more genuinely historical form of revisionism. These would include John H. Brooke, Reijer Hooykaas, Eugene Klaaren, J.E. McGuire, Francis Oakley and Margaret J. Osler—individuals who do not share a common ideology (so far as I know), and therefore cannot be accused as a group of engaging in thinly veiled apologetics.[16] Taken together, their work goes a long way toward bridging the gap between Foster's abstract assumptions and the complex historical reality of early modern science.

It is from my own work, however, that I will offer some representative examples of how different theologies of creation were used to undergird different conceptions of scientific knowledge by early modern natural philosophers. Elsewhere in this collection I have given a lengthy account of the role of voluntarist theological assumptions in the thought of Isaac Newton (1642-1727), which readers are encouraged to study as a supplement to this essay.[17] Here I will consider, much more briefly, three other major natural philosophers of the seventeenth century: Galileo Galilei (1564-1642), René Descartes (1596-1650) and Robert Boyle (1627-1691).[18] These case studies, which should be viewed as vignettes taken from much longer, more systematic essays,[19] will be used to test the revisionist thesis by sketching the answers to three main questions: How was God's relationship to the human mind understood? How was God's relationship to nature understood? What overall view of the nature of scientific knowledge was proclaimed? Although I will not always quote authors directly, I have been careful to rely on their explicit statements about God, and to avoid arguing (as Foster often does) by implication and inference. As we will see, Foster is correct to suggest that theological assumptions are closely associated with conceptions of scientific knowledge. However, Foster's belief that Christian theology actually caused the rise of modern science turns out to be as much an oversimplification of the actual situation as the belief that warfare between religion and science is inevitable.

THREE CASE STUDIES, BRIEFLY CONSIDERED

Apart from his famous letter to Christina de Médici, written in response to a specific threat to freedom of thought, Galileo Galilei never writes extensively or systematically about any theological subject. Nevertheless, he makes enough explicit theological statements that it is possible for us to assemble a picture of his theology of creation that is consistent with everything else we know about him. In confronting the Aristotelian philosophy of nature that dominated the universities of his age, Galileo is forced to justify his alternative view that physics is essentially mathematics applied to matter, and the justification he provides is couched in the language and concepts of theology. Galileo's ideal of a mathematical, a priori science of nature is grounded explicitly in a rationalistic, highly Platonic, understanding of God's relation to created objects and to created minds.

This is not to deny that Galileo performed many experiments, which certainly helped him formulate propositions about nature; I mean simply that his *ideal* of scientific knowledge is that of mathematical demonstration, which he believes to be the only way to achieve certainty. For Galileo, no less than for Aristotle, certainty rather than probability is the ultimate goal of science.[20] Galileo virtually equates science, understood as true and necessary knowledge of nature, with mathematics, true and necessary demonstration. He seeks not to persuade opponents by probable arguments, but to prove his conclusion "by necessary demonstrations from their primary and unquestionable foundations."[21] Natural truths, he says, "must follow necessarily, in such a way that it would be impossible for them to take place in any other manner," for "just as there is no middle ground between truth and falsity in physical things, so in rigorous proofs one must either establish his point beyond any doubt or else beg the question inexcusably."[22] It was the failure of other natural philosophers to do just this that Galileo was wont to criticize most. Thus he says of William Gilbert's work on magnetism, that his reasons "lack that force which must unquestionably be present in those adduced as necessary and eternal scientific conclusions."[23]

Galileo realizes that a truly demonstrative science would be possible only if God, as Author of the great book of the universe, guaranteed that nature displayed the same necessity as the mathematical language in which it was written, and further guaranteed that the human mind was capable of reading that language. He claims nothing less. As "the obedient executrix of God's commands," nature is "inexorable and immutable," never transgressing "the laws imposed upon her."[24] Since he takes "matter to be inalterable—that is, always the same," it is evident that for any "eternal and necessary property, purely mathematical demonstrations can be produced."[25] Now Galileo is not a pure Platonist: unlike Plato, he thought we could attain true knowledge of material bodies. But he confesses that he is not "far from being of the same opinion" with Plato, who "admired the human understanding and believed it to partake of divinity simply because it understood the nature of numbers."[26] Within the range of mathematical demonstration, Galileo claims, human "knowledge equals the Divine in objective certainty, for here it succeeds in understanding necessity, beyond which there can be no greater sureness."[27]

Galileo does place limits upon our ability to plumb the depths of nature, but these arise from our status as creatures with finite minds rather than from an exercise of divine freedom. Thus we can not know infinitely many truths, Galileo admits, for we are not God. We could know discursively by reason some of the truths that God knows instantly by intuition, and here our knowledge is equal to God's.[28] But our limited capacity to comprehend the subtle mathematics of infinity, which God has sometimes used in making the world, means there are some things about nature we will never fully understand. In an interesting letter to Gallanzone Gallanzoni, secretary to Cardinal François de Joyeuse (1559-1615), Galileo identifies three classes of proportions—the perfect, between proximate numbers; the less perfect, between "more remote prime numbers"; and the imperfect, between incommensurables, such as ratios of square roots to one another. If it were up to us to arrange the motions of the celestial bodies, Galileo says, "we should have to rely on proportions of the first type, which are the most rational." But God, "not bothering about symmetries that man can understand," had used imperfect proportions, which were not only "irrational, but totally inaccessible to our intelligence." Likewise, Galileo continues, if a famous architect were to distribute the fixed stars throughout the vault of heaven, he would employ regular geometric figures and familiar ratios. But God, "by apparently scattering them at random, impresses us as having arranged them without heeding any rules or any demands of symmetry and elegance."[29] Nor can we determine whether the world is finite or infinite in size, Galileo tells the philosopher Fortunio Liceti (1577-1657) late in his life. Galileo leans toward an infinite world simply because "I feel that my incapacity to comprehend might more properly be referred to incomprehensible infinity, rather than to finiteness, in which no principle of incomprehensibility is required." Ultimately, he confesses, this is one of those questions that is "happily inexplicable to human reason, and similar perchance to predestination, free will, and others in which only Holy Writ and divine revelation can give an answer to our reverent remarks."[30]

For all of Galileo's willingness to surrender infinity to God's unlimited mind, God's will has almost no place in Galilean natural philosophy, not even in thought experiments, the traditional vehicle for the consideration of supernatural possibilities. For example, Galileo twice suggests the experiment of dropping a body into a tunnel through the center of the earth, precisely the sort of situation in which many medieval natural philosophers would have invoked God's absolute power to do such a thing. Neither time did Galileo suggest an agent.[31] Equally he refuses to accept without sarcasm supernatural possibilities proposed by others.[32] The principal role for God's power in Galileo's natural philosophy is implicit rather than explicit: to ensure that God's geometrical thoughts would be completely realized in the objects of his making, so that even the accidental complexities of real bodies would be subject to the exacting scrutiny of geometry.[33] Although nature is not simple, it is nevertheless almost wholly explicable in mathematical terms. For Galileo, in the words of one modern scholar, there is "no intractable surd in the things of nature which defies rationalization."[34] And since the world is mathematical to the core, our failure completely to know God's works is attributed much more to the defect of a finite human understanding than to the exercise of divine freedom. Thus Galileo's theology of creation, like his conception of science, was essentially rationalistic.

The same could be said of René Descartes, who ironically uses an extreme form of voluntarism as a tool to help establish a rationalistic natural philosophy. Emphasizing God's utter transcendence, Descartes denies that any limits can be placed on God's power to create a boundless universe[35] or eternal truths incomprehensibly different from those that the human mind perceives as necessary.[36] In this he goes even further than any medieval nominalist I am aware of. However, Descartes's God can not change his mind once he has chosen which truths to create, because he is perfect, and his will and intellect are one. Thus if we can somehow gain knowledge of these truths, that knowledge will be permanent and necessary, not contingent. We can indeed get such knowledge, Descartes believes, for God has given us certain innate truths. Here Descartes retreats from the radical voluntarism he expresses elsewhere. A perfect God can not deceive us by implanting in our souls seeds of error rather than seeds of truth. Thus the human mind becomes the touchstone for created reality: what it can not comprehend, God obviously has not chosen to create.[37]

Thus Descartes uses the voluntarist notion of a free Creator as a *reductio ad absurdum*, in a way that closely parallels the process of methodic doubt for which he is so well known. The evil genius who haunts Descartes's sleep of reason is exactly like the radically free God who can create arbitrary and continually changing truths. Either one will undermine the possibility of absolute certainty, which Descartes can not allow. Both the evil genius and the nightmare—God has to be dismissed by an act of faith in the veracity of a more reasonable God who can not change his mind or deceive his human creations.

At the same time, God's freedom to employ his absolute power was not altogether denied. Initially to determine which laws to make true and which mechanisms to place in nature remains the privilege of the divine will, not the human mind; the results of that determination are shared only partially. The general fabric of the world follows from the first principles of Cartesian physics (the laws of motion, and the equivalence of matter and extension), which are fully revealed to us by the light of reason and can not be otherwise, so that there can only be one *kind* of world.[38] But within this world, God could have placed an infinity of particular things, unfettered by the necessity to choose one or another to instantiate.[39] And the general principles by which Descartes seeks to explain these contingent things are "so vast and so fertile," "so simple and so general," that a particular phenomenon can be explained in any number of ways.[40] God could have chosen to employ any one of an infinite number of possible mechanisms to produce a given phenomenon. Which one he has actually chosen can be found only by experimentation.[41] The human mind can not separate reality from possibility, actuality from contingency, without an appeal to experience.

Descartes therefore sees the impossibility of creating a purely a priori physics all the way down to the last detail. In a letter to the Minim friar Marin Mersenne (1588-1648), which is strikingly similar to Galileo's letter to Gallenzoni, Descartes tells of how he has sought to find "the cause of the position of each fixed star." Though they seem "very irregularly distributed in various places in the heavens," Descartes believes there is "a natural order among them which is regular and determinate." To discover this would be

> the key and foundation of the highest and most perfect science of material things which men can ever attain. For if we possessed it we could discover *a priori* all the different forms and essences of terrestrial bodies, whereas without it we have to content ourselves with guessing them *a posteriori* from their effects.

But such a discovery would depend on having an exhaustive knowledge of particular facts about the stars, which Descartes despaired of gaining. "The science I describe," he has to admit, "is beyond the reach of the human mind; and yet I am so foolish that I cannot help dreaming of it though I know this will only make me waste my time...."[42]

But Descartes's dream of an a priori science was never really given up. In spite of an unmistakable element of voluntarism in his theology of nature, Descartes's *ideal* of science remains (like Galileo's) essentially rationalistic. Though God was free to produce the particulars of nature in a variety of ways, the fundamental laws of nature necessarily conform absolutely to the innate truths implanted in our souls, or else God would be deceiving us. Grounded in such a theology of creation, Cartesian natural philosophy is largely rationalistic; experience is required to augment pure reason precisely and only where the vestiges of divine will remain.

Robert Boyle agrees with Descartes in identifying the basic laws of mathematics and logic as eternal truths which function as the foundation of all knowledge, including theology as well as science. He also affirms that God, as "the author of our reason, cannot be supposed to oblige us to believe contradictions," and that God's veracity and boundless knowledge prevented him from deceiving us.[43] But where Descartes also embraces the proposition that all things perceived clearly and distinctly are true—including the fundamental laws of nature, which could therefore be known from pure reason—Boyle places the foil of divine freedom abruptly in the path of a rationalistic science. We "mistake and flatter human nature too much," he says, "when we think our faculties of understanding so unlimited,...as many philosophers seem to suppose." Boyle emphasizes that we are created, "as it pleased the almighty and most free author of our nature to make us." Our mental abilities were "proportionable to God's designs in creating us, and therefore may probably be supposed not to be capable of reaching to all kinds...of truths, many of which may be unnecessary for us to know here...."[44] As "purblind mortals," we can be only "incompetent judges" of God's power, which can "justly be supposed to reach farther than our limited intellects can comprehend; or for that reason, without a saucy rashness, can presume to bound."[45] "For if we believe God to be the author of things," he continues later, "it is rational to conceive that he may have made them commensurate, rather to his own designs in them, than to the notions we men may best be able to frame of them." The world was made before human beings, who had not been consulted in its construction! The author of nature "made things in such a manner as he was pleased to think fit," leaving people "to speculate as well as they could upon those corporeal, as well as other things."[46]

Boyle's concept of a scientific law is wholly consistent with this voluntarist attitude toward God's creative activity. The laws of nature, no less than the rest of the creation, are imposed upon matter by the free choice of a sovereign God. Natural laws, as Boyle puts it, "did not necessarily spring from the nature of matter, but depended on the will of the divine author of things...."[47] Boyle does not believe that matter is wholly passive *mechanically* (entirely without powers of its own), as some scholars have thought,[48] but he absolutely believes that matter is wholly passive *ontologically*. He puts this clearly in an unpublished manuscript from the 1680s:

> The Primordial system of the universe, or the great & Original fabrick of the world; was as to us arbitrarily establisht by God. Not that he created things without accompanying, & as it were regulating, his omnipotence, by his boundless wisdome; & consequently did nothing without weighing reasons: but because those reasons are a priori [crossed out: unsearchable] undiscoverable by us: such as are the number of the fixt stars, the colocation as well as number of the planetary globes, the lines & period of their motion,...the bigness, shapes, & differing longevity of Living creatures; & many other particulars: of which the onely Reason we can assign, is that it pleasd God at the beginning of things, to give the world & also its parts the disposition. (This may also be applied to the states of bodys & the rules of motion.)[49]

Thus we find Boyle saying of other worlds, that they "may be framed and managed in a manner quite differing from what is observed in that part of the universe, that is known to us." The kind of matter, the laws of motion, and the living creatures might be highly unlike those in our own world.[50] Similarly, in the new heaven and earth that God will some day substitute for this one, "the primordial frames of things, and the laws of motion, and consequently, the nature of things corporeal, may be very differing from those that obtain in the present worlds."[51] As he says elsewhere, only "God knows particularly, both why and how the universal matter was first contrived into this admirable universe, rather than a world of any other of the numberless constructions he could have given it...."[52]

Surely this is a voluntarist's world. Because the Creator has worked and continues to work in accordance with his free will and not out of necessity, the order of nature is contingent and could not be known a priori. Boyle says as much when he describes physical laws as "collected or emergent" truths "gathered from the settled phaenomena of nature," not "axioms metaphysical, or universal, that hold in all cases without reservation."[53] Boyle's theology of creation is markedly more voluntaristic than that of Galileo or Descartes, precisely because he had much less confidence in fallen human minds, and places more emphasis on God's freedom and absolute power. The key word here is "emphasis." Certainly Galileo and Descartes *believed* in divine freedom

and power—what seventeenth-century Christian did not? The important
question is, did they *emphasize* them to the extent that Boyle did? I
think not. Where Galileo and Descartes reluctantly conceded that some
things lay beyond the power of finite minds to know, Boyle practically
glories in our status as "purblind mortals," and constantly reminds his
readers that it would be "saucy rashness" for them to "presume to
bound" God's freedom to make the world as he saw fit.

CONCLUSIONS

What can now be said about the relation between theology of creation
and philosophy of nature in the seventeenth century? At very least, we
have seen that theological presuppositions are closely allied with
conceptions of scientific knowledge, and in just the way that Foster
suggests. Galileo's lack of emphasis on divine freedom is mirrored in
his a priori attitude toward scientific knowledge. Descartes's belief that
God cannot deceive us leads him to claim that the first principles of
physics can be discovered by pure reason. For Boyle, however, an
emphasis on divine will goes hand in hand with a commitment to the
primacy of phenomena.[54]
  But are these parallel emphases any more than just parallel
emphases, as interesting as it may be to discover them? No single
answer is likely to satisfy everyone. If one believes (as I do) that
thoughtful people like those studied here tend to strive for consistency
between various parts of their minds, then these parallel emphases are,
in and of themselves, evidence that a two-way conversation between
science and theology has taken place, leading toward a unified vision
of reality. As Koyré once observed, "The God of a philosopher and his
world are correlated."[55] Others might call for more words of the
conversation, as it were, spelling out precisely when a given thinker
came to believe precisely what about God or the world for a specific
reason. For them, a definitive answer would require detailed
biographical studies of a sort which have not yet been done for any of
the three figures here.[56] Still others might claim that theological
language is never able to do any more than give ultimate expression to
beliefs one has already formed on more purely philosophical or
scientific grounds. Adherents of the conflict model of theology and
science, for example, would find it hard to admit even the possibility
that theology has helped shape the modern conception of scientific
knowledge: for them, theology can only retard genuine science, not
inform it. The statements quoted in this paper are not likely to convince
them that a genuine dialogue has taken place; nor can I imagine what
sort of account *would* be convincing, given their particular bias.

There remains yet another group: those who are convinced, as Foster was, that Christian theology actually *caused* modern science, insofar as it led early modern thinkers to break with Greek notions of science in direct response to biblical teaching about the contingency of created nature. I reject such a claim—not because I doubt that theology influenced science in this way, but because I do not accept the narrow definition of modern science that is implicit in the claim. During what is often called the scientific revolution, several fundamental changes took place, each the result of a variety of factors. To give just one example, the change in world picture from geocentrism to heliocentrism certainly shows the influence of Neoplatonism on men like Copernicus, Kepler and Galileo; but voluntarist theology did not contribute anything of importance to this fundamental aspect of the scientific revolution.

What voluntarism *did* affect was the debate about what sort of knowledge the new science ought primarily to be: necessary truths demonstrable from pure reason, or contingent truths emergent from phenomena. But even here, theological factors were not the only ones of importance. When Renaissance humanists rediscovered the ancient sceptics, they gave European thinkers a foil to rationalism so powerful that Descartes made it the prime target of his program to reconstruct knowledge. His countryman Pierre Gassendi, however, made a mitigated Pyrrhonian scepticism the epistemological basis for his version of the mechanical philosophy, which was itself deeply influential on both Robert Boyle and Isaac Newton.[57] To bring the argument around full circle, we note that Reformation debates about the sources of religious authority were in turn a principal factor in creating a sceptical climate in the first place.[58] The real historical picture is complex: science, philosophy and theology are inextricably intertwined. To single out one factor as the sole cause is to do violence to the actual situation.[59]

My overall conclusion, then, is that seventeenth-century natural philosophers operated within a theological framework that they did not ignore when they created their new view of scientific knowledge. Their assumptions about God's relation to created minds and the created order, when made explicit, show that theological presuppositions were in fact closely tied to presuppositions about the nature of scientific knowledge. We need, however, to be careful not to overstate this conclusion. Voluntarist theology neither "caused" modern science nor, in and of itself, did it lead to a particular kind of science. It was rather one factor, albeit a very important one, in giving modern science its strong empirical bent.

NOTES

1. From the preface to R.S. Westfall, *Science and Religion in Seventeenth-Century England* (New Haven: Yale University Press, 1973).

2. See the preface to R.K. Merton, *Science, Technology and Society in Seventeenth-Century England*, 2nd ed. (New York: Harper and Row Publishers, 1970), xv-xvi.

3. Merton, xvi.

4. For spirited, well-documented essays on the damage done to the landscape of historiography by the conflict thesis, see C.A. Russell, "Some Approaches to the History of Science," in *The "Conflict Thesis" and Cosmology*, edited by C.A. Russell, R. Hooykaas and D.C. Goodman (Milton Keynes: Open University Press, 1974); J.R. Moore, *The Post-Darwinian Controversies: A Study of the Protestant Struggle to Come to Terms with Darwin in Great Britain and America, 1870-1900* (Cambridge: Cambridge University Press, 1979); and D.C. Lindberg and R.L. Numbers, "Beyond War and Peace: A Reappraisal of the Encounter between Christianity and Science," *Church History* 55 (1986): 338-54.

5. Merton told me that Foster's work was unknown to him until he read a draft of an earlier version of this essay in 1988.

6. See M.B. Foster, "The Christian Doctrine of Creation and the Rise of Modern Natural Science," *Mind* 43 (1934): 446-68; and M.B. Foster, "Christian Theology and Modern Science of Nature," *Mind* 44 (1935): 439-66 and 45 (1936): 1-27. For a biography of Foster, see C. Wybrow, "Introduction: The Life and Work of Michael Beresford Foster," in *Creation, Nature, and Political Order in the Philosophy of Michael Foster (1903-1959): The Classic "Mind" Articles and Others, with Modern Critical Essays*, edited by C. Wybrow (Lewiston and Lampeter, Wales: Edwin Mellen Press, 1992), 3-44, a thoroughly researched essay that goes well beyond the limited information in [M.B. Foster,] Obituary in *London Times* (October 16, 1959): 15; [M.B. Foster,] Obituary in *Manchester Guardian* (October 16, 1959): 19; and Memorial Sermon for M.B. Foster, by V.A. Demant, *Christian Scholar* 43 (1960): 3-7; and includes a splendid overview of the whole range of his thought. I offer more extensive comments on Wybrow's collection of papers in E.B. Davis, "Book Review of *Creation, Nature, and Political Order in the Philosophy of Michael Foster (1903-1959): The Classic "Mind" Articles and Others, with Modern Critical Essays*, edited by C. Wybrow," *Isis* 85 (1994), 127-29. Reprinted with editorial changes and additions as "Christianity and Early Modern Science: Beyond War and Peace?" *Perspectives on Science and Christian Faith* 46 (1994): 133-35. For a critical analysis of the internal logic of Foster's argument, see I. Weeks and S. Jacobs, "Theological and Philosophical Presuppositions of Ancient and Modern Science: A Critical Analysis of Foster's Account," in *Creation, Nature, and Political Order in the*

Philosophy of Michael Foster (1903-1959): The Classic "Mind" Articles and Others, with Modern Critical Essays, edited by C. Wybrow (Lewiston and Lampeter, Wales: Edwin Mellen Press, 1992), 255-68.

7. On the origins and uses of this distinction, see W.J. Courtenay, "Nominalism and Late Medieval Religion," in *The Pursuit of Holiness in Late Medieval and Renaissance Religion*, edited by C. Trinkaus and H. Oberman (Leiden: E.J. Brill, 1974), 37-43; quoting 37. Technically, *potentia absoluta* (God's power considered absolutely) and *potentia ordinata* (God's power as ordained through his will for the creation) are not identical with will and reason *per se*, but they reflect the same dialogue between freedom and order.

8. These terms are used by Foster, and in much of the secondary literature cited in this essay.

9. Quotations taken from Foster, "Christian Doctrine of Creation," 465; and Foster, "Christian Theology and Modern Science," 1, 10, and n.5, respectively.

10. His papers are replete with statements that would support this, although it is rarely stated directly. An exception is found on page 11 of the 1936 article. If it should happen that certain aspects of Cartesian philosophy are "compatible with a theology which has been found to be unorthodox, and therefore unchristian," Foster says, "the whole contention that there is a necessary implication between Christian theology and modern natural science would be imperilled."

11. R. Gruner, "Science, Nature, and Christianity," *Journal of Theological Studies* 26 (1975): 56. Gruner's article has recently been reprinted in *Creation, Nature, and Political Order in the Philosophy of Michael Foster (1903-1959): The Classic "Mind" Articles and Others, with Modern Critical Essays*, edited by C. Wybrow (Lewiston and Lampeter, Wales: Edwin Mellen Press, 1992), 213-243.

12. As Cartmill observes, "The Baconian tradition no longer has any literary champions. There is a growing consensus—outside the scientific research community, at any rate—that limits must be placed on human power over nature" (M. Cartmill, *A View to a Death in the Morning: Hunting and Nature through History* [Cambridge: Harvard University Press, 1993], 215).

13. For example, his point that modern science did not appear for more than fifteen hundred years after the origin of Christianity, which seems to counter the claim that Christianity was so favorable to it.

14. Gruner, 65.

15. Whether one ought to take the historical argument further, by heaping praise on Christianity for its role in helping to shape modern science, is another question entirely. No doubt some people would be impressed with such an argument. But others might be just as tempted to use alternative historical

arguments to blame Christianity for such things as anti-Semitism, the torture and execution of witches or (as Gruner fears) environmental exploitation.

16. For representative works by these authors, see the bibliography. It is impossible to cite more than just a few of the numerous books and articles related to the Foster thesis. I have chosen a group of scholars whose works have been, in my judgment, among the most influential on others. It is worth noting that at least two members of this group did not know of Foster's work until their own work was quite advanced. In commenting on an earlier version of this paper at another conference, Reijer Hooykaas stated as much, adding that he learned about Foster from the late British cyberneticist Donald M. MacKay. (*Religion and the Rise of Modern Science* is dedicated to MacKay's mother.) Francis Oakley first saw Foster's articles in 1959-60, after his ideas were "essentially in place and complete." See the afterword to the reprint of his article in *Creation, Nature, and Political Order in the Philosophy of Michael Foster (1903-1959): The Classic "Mind" Articles and Others, with Modern Critical Essays*, edited by C. Wybrow (Lewiston and Lampeter, Wales: Edwin Mellen Press, 1992), 209-11.

17. E.B. Davis, "Newton's Rejection of the 'Newtonian World View,'" *Fides et Historia* 22 (1991): 6-20. Reprinted in *Science and Christian Belief* 3 (1991): 103-17. Also reprinted with minor additions as "Newton's Rejection of the 'Newtonian Worldview ': The Role of Divine Will in Newton's Natural Philosophy," in *Facets of Faith and Science. Volume 3: The Role of Beliefs in the Natural Sciences*, edited by J.M. van der Meer (Lanham: The Pascal Centre for Advanced Studies in Faith and Science/University Press of America, 1996).

18. I am convinced that the best test of the Foster thesis is to apply it to some of those who were regarded by their contemporaries as the leading natural philosophers of the age. It is not "whiggish" or unhistorical to say this: it is one thing to select a group of individuals based solely upon our perception *today* that their ideas led to where we are *now*; but it is another thing entirely to select thinkers who were seen *at the time* to be the thinkers whose ideas were of the greatest significance *for that time*. Each of the three people chosen here was viewed in that way; the fact that we also tend to see them as having led to where we are now does nothing to alter their importance in the context of the scientific revolution, or to diminish their suitability as subjects of this study.

19. It is assumed that the present essay is aimed at a broad scholarly audience, people who are likely to have less interest in the details and more in the overall argument and conclusions. For more comprehensive treatments, readers are invited to consult E.B. Davis, *Creation, Contingency, and Early Modern Science: The Impact of Voluntaristic Theology on Seventeenth-Century Natural Philosophy* (unpublished Ph.D. dissertation, University of Indiana, 1984) for Galileo and Boyle; and for Descartes: E.B. Davis, "God, Man, and

Nature: The Problem of Creation in Cartesian Thought," *Scottish Journal of Theology* 44 (1991): 325-48.

20. For accounts of Galileo's rationalism with which I wholeheartedly agree, see E. McMullin, "The Conception of Science in Galileo's Work," in *New Perspectives on Galileo*, edited by R.E. Butts and J.C. Pitt (Dordrecht: D. Reidel Publishing Company, 1978), 209-57; and A.C. Crombie, "The Sources of Galileo's Early Natural Philosophy," in *Reason, Experiment, and Mysticism in the Scientific Revolution*, edited by M.L. Righini Bonelli and W.R. Shea (New York: Science History Publications, 1975), 157-75, a study of Galileo's early writings that stresses the degree to which he accepts the Aristotelian notion of scientific knowledge as certain knowledge.

21. G. Galilei, *Discourses on Two New Sciences*, translated by S. Drake (Madison: University of Wisconsin Press, 1974), 15.

22. Taken (respectively) from G. Galilei, *Dialogue Concerning the Two Chief World Systems—Ptolemaic and Copernican*, translated by S. Drake (Berkeley: University of California Press, 1953), 424; and G. Galilei, *The Assayer*, in *The Controversy on the Comets of 1618*, edited by S. Drake and C.D. O'Malley (Philadelphia: University of Pennsylvania Press, 1960), 252. Cf. Galilei, *Dialogue Concerning the Two Chief World Systems*, 157ff.

23. Galilei, *Dialogue Concerning the Two Chief World Systems*, 406.

24. G. Galilei, *Letter to the Grand Duchess Christina*, in *Discoveries and Opinions of Galileo*, edited by S. Drake (Garden City: Doubleday, 1957), 182. Cf. the similar passage from his *Letters on Sunspots* on 136 in the same volume; and his letter to Elia Diodati, quoted by T.P. McTighe, "Galileo's Platonism: A Reconsideration," in *Galileo: Man of Science*, edited by E. McMullin (New York: Basic Books, 1967), 375.

25. Galilei, *Discourses on Two New Sciences*, 13.

26. Galilei, *Dialogue Concerning the Two Chief World Systems*, 11f.

27. Galilei, *Dialogue Concerning the Two Chief World Systems*, 103f.

28. Galilei, *Dialogue Concerning the Two Chief World Systems*, 103f.

29. The letter is quoted at length in M. Clavelin, *The Natural Philosophy of Galileo: Essay on the Origins and Formation of Classical Mechanics*, translated by A.J. Pomerans (Cambridge: MIT Press, 1974), 447f.

30. Quoted by A. Koyré, *From the Closed World to the Infinite Universe* (Baltimore: Johns Hopkins University Press, 1957), 98. As Koyré notes just before this, Galileo flatly denies the infinite size of the world in the *Dialogues*, but this may have been done simply to get the book approved.

31. Galilei, *Dialogue Concerning the Two Chief World Systems*, 22 and 236.

32. For more on this aspect of his writing, see Davis, *Creation, Contingency, and Early Modern Science*, 54-58.

33. This is how I understand passages such as Galilei, *Dialogue Concerning the Two Chief World Systems*, 203-8; and Galilei, *Discourses on Two New Sciences*, 12-13.

34. McTighe, 369.

35. See, among other places, a remark in the *Excerpta ex Cartesio: MS de Leibniz*, in R. Descartes (1897-1913), *Oeuvres de Descartes*, edited by C. Adam and P. Tannery, 13 vols. (Paris: Leopold Cerf), 11:656.

36. Descartes discusses the eternal truths several times in his correspondence, and in a few other places. A nice example relevant to my point here is found at the end of his letter to Arnauld of 29 July 1648; see Descartes, 5:223.

37. For a more extended discussion of these points, see Davis, "God, Man, and Nature," 328-35.

38. It is clear that Descartes thought that these principles were evident from metaphysical considerations. The identity of matter and extension follow from his inability to conceive them clearly as separate things; and the laws of motion follow from the immutability of God. See the second part of Descartes, *Principles of Philosophy*, 8A:40-79.

39. This is a central point of the famous passage on scientific method in the sixth part of the Descartes, *Discourse on Method*, 6:63-5.

40. Descartes, *Principles of Philosophy*, iii, 4, 8A:81f; and the sixth part of his *Discourse on Method*, 6:64.

41. Descartes, *Principles of Philosophy*, iii, 46, 8A:100f.

42. Quoting from the translation by A. Kenny, *Descartes: Philosophical Letters* (Oxford: Oxford University Press, 1970), 23f. Had Descartes somehow managed to gain the exhaustive knowledge of particulars he lacked, it would only have shown him once again the necessity of having empirical knowledge for answering certain questions.

43. See *The Christian Virtuoso*, in R. Boyle, *The Works of the Honourable Robert Boyle*, edited by T. Birch, 6 vols. (London: Millar, 1772), 6:709-12 and 5:529.

44. *A Discourse of Things above Reason*, in Boyle, 5:410. Boyle's position on these sorts of truths is fully explicated in J.W. Wojcik, "The Theological Context of Boyle's *Things Above Reason*," in *Robert Boyle Reconsidered*, edited by M. Hunter (Cambridge: Cambridge University Press, 1994), 139-155.

45. From *The Christian Virtuoso*, in Boyle, 6:676f.

46. From *The Christian Virtuoso*, in Boyle, 6:694.

47. *The Christian Virtuoso*, in Boyle, 5:521.

48. See, e.g., J.E. McGuire, "Boyle's Conception of Nature," *Journal of the History of Ideas* 33 (1972): 523-42; and G.B. Deason, "Reformation Theology and the Mechanistic Conception of Nature," in *God and Nature: Historical Essays on the Encounter between Christianity and Science*, edited by D.C. Lindberg and R.L. Numbers (Berkeley: University of California Press, 1986), 167-91. For a useful corrective, see T. Shanahan, "God and Nature in the Thought of Robert Boyle," *Journal of the History of Philosophy* 26 (1988): 547-69.

49. Royal Society, Boyle MS. 185, fol. 29, quoted with permission of the President and Fellows of the Royal Society. The manuscript, in the hand of Boyle's amanuensis Thomas Smith, is from an intended appendix for *A Disquisition about the Final Causes of Natural Things*, in Boyle, 5:392-444.

50. Boyle, *Of the High Veneration Man's Intellect Owes to God*, in Boyle, 5:139.

51. *The Christian Virtuoso*, in Boyle, 6:788f. The importance of statements about other worlds, including the new one God has not yet made, for studies of the limits of scientific knowledge in the seventeenth century cannot be underestimated.

52. *High Veneration*, in Boyle, 5:149f.

53. *Things above Reason*, in Boyle, 4:462f.

54. As I have shown elsewhere (Davis, "Newton's Rejection of the 'Newtonian World View'"), divine will was also primary for Isaac Newton, who held similar views about the laws of nature and our inability to discover them from pure reason. An emphasis on the dominion of an ever active, omnipresent, and free creator lay at the heart of Newton's natural philosophy.

55. Koyré, 100.

56. Existing biographical studies explore both scientific and theological beliefs, but do not focus on the questions asked in this essay. Although I have made some attempt to construct such accounts in Davis, "Newton's Rejection of the 'Newtonian World View'"; and Davis, "God, Man, and Nature"; mostly I have been concerned simply to recover what each of these thinkers actually says about God, nature and the human mind. Placing their statements more fully within the context of intellectual biography remains to be done.

57. On this particular point, see M.J. Osler, "The Intellectual Sources of Robert Boyle's Philosophy of Nature: Gassendi's Voluntarism and Boyle's Physico-Theological Project," in *Philosophy, Science, and Religion in England, 1640-1700*, edited by R. Kroll, R. Ashcraft and P. Zagorin (Cambridge: Cambridge University Press, 1992), 178-198.

58. On Descartes, Gassendi, the revival of scepticism and the role of the Reformation, see R. Popkin, *A History of Scepticism, From Erasmus to Spinoza*, rev. ed. (Berkeley: University of California Press, 1979). The fact that Gassendi was a Catholic voluntarist is worthy of note. Voluntarism was not, as this paper might implicitly suggest, a Protestant theology. There were Catholic voluntarists like Gassendi, and Protestant rationalists like Gottfried Leibniz.

59. I am happy to echo John Brooke's central theme: "Serious scholarship in the history of science has revealed so extraordinarily rich and complex a relationship between science and religion in the past that general theses are difficult to sustain. The real lesson turns out to be the complexity" (J.H. Brooke, *Science and Religion: Some Historical Perspectives* [Cambridge: Cambridge University Press, 1991], 5).

Part 2:
Beliefs in the Biological Sciences

10

The Concept of the
"Open System"—
Another Machine Metaphor for the
Organism?[1]

Sytse Strijbos

THE ORGANISM AS A LIVING MACHINE

The machine as a metaphor for the organism is a classical topic in discussions of the foundations of biology and the medical sciences.[2] Since the beginnings of modern science, the model of the living machine has been adopted for the organism. Various versions of this model appeared, depending on the state of technology. The construction of theory in the field of biology—and not there alone, for that matter, but also in the human and social sciences—was thereby kept closely linked to the continuing development of technology. Thus, in the seventeenth century, models were taken from the construction and functioning of the mechanical clock. Later, in the eighteenth and nineteenth centuries, when thermodynamics and the steam engine were introduced, the model of the steam engine became popular, as Russelman has shown, for psychology and psychiatry.[3] And in the present century, since the rise of information theory and of cybernetics and computer technology in the forties, the metaphor of the computer has gained rapid entry into countless fields.

How are we to regard the metaphorical function of technology? A familiar interpretation was presented recently by David Bolter in a study about artificial intelligence and the computer metaphor.[4] Attempts to explain the natural world through models of technical constructions are not, so this writer says, an innovation of recent centuries. Constantly engaged in drawing anew the line between nature and culture, humanity is permanently disposed to explain the former in terms of the latter, to view the natural world through the glasses of his own human-made environment. By way of illustration, Bolter recalls that the Greek philosophers already used comparisons with the crafts of the potter and woodworker to explain the creation of the universe. Likewise, in the Middle Ages, the weight-driven clock provided a new metaphor for many matters.

All technologies and tools have the potential to become defining technological material. "A defining technology," according to Bolter,

> resembles a magnifying glass, which collects and focuses seemingly disparate ideas in a culture into one bright, sometimes piercing ray. Technology does not call forth major cultural changes by itself, but it does bring ideas into a new focus by explaining or exemplifying them in new ways to larger audiences.[5]

In reality, in any given period only one or a few technological inventions or artifacts prevail and gain the status of a defining technology. Bolter identifies the clock, the steam engine and the computer as metaphors in successive stages of the history of scientific thought.

Essential in Bolter's perspective on the metaphorical function of technology is that technology by nature must imply a redefinition of the relation of humans to nature. Human identity and place in the world are in Bolter's view matters of self-determination. Bolter considers the human person to be a self-defining subject. It is this subjectivistic philosophy in the line of Descartes that underlies the effort to make technology the model for natural, unmade reality, including humanity itself. Thus the technological metaphor is not necessarily *given* with the development of technology or with being human; rather, it is a product of the prevailing view of humanity and the world, the view that also conditions the background of Bolter's reflections. The role of technology as a metaphor in scientific theories can be identified as a fruit of the technical world picture that dominates modern culture.[6]

What are the implications of these considerations about the machine metaphor in science for the view of the organism as a living machine? Bolter's interpretation of the function of technology as a metaphor implies a justification of the view of the organism as a living machine. For in this view it belongs to human nature to constantly redefine the natural world and human relation to it in terms of self-conceived technical constructions. My perspective, in contrast, subjects the conception of the animate mechanism to fundamental criticism. This conception, according to the criticism, is determined by an underlying view of the human person and the world that is far from self-evident and that places far too much stock in the possibilities of theoretical thought in general, and of machine models, in particular.

Von Bertalanffy criticizes various machine models of the organism and attempts to offer, in the conception of the organism as an open system, a scientific alternative. The first question we must investigate concerns the role played by "metaphysical beliefs" in von Bertalanffy's criticism of the machine metaphor and in his new conception of the organism. A second fundamental question, which follows from the first, is: did von Bertalanffy succeed in freeing himself from the tradition of mechanical thinking in biology, or does his "open system" represents a new machine metaphor?

THE ORGANISM AS AN OPEN SYSTEM

While in our everyday experience we have (as a rule) no difficulty distinguishing between animate and inanimate objects, the theoretical interpretation of the difference is less simple. Since the rise of modern science a model for this difference has been that of the living machine with its variants as described above. And although von Bertalanffy acknowledges that this model is not without merit and affords insight into some phenomena, he also recognizes its difficulties and limitations. Von Bertalanffy identifies at least three problems.

In the first place, there is the unresolved problem of the origin of the machine. Clocks, steam engines and computers do not grow of themselves, but are human creations. Where does the infinitely more complex living machine come from? Secondly, there is the capacity of organisms to recover from various disturbances (regulation) that impinge upon them from their surroundings. Most important to von Bertalanffy is a third unresolved problem in the conception of the organism as a living machine. A characteristic of the living organism is that it maintains itself in a continuous exchange of components. A machine lacks the capacity to do so. And in von Bertalanffy's reflections, it follows that machinelike structures cannot form the ultimate basis of living phenomena:

> We have, as it were, a machine composed of fuel spending itself continually and yet maintaining itself. Such machines do not exist in present-day technology. In other words: A machinelike structure of the organism cannot be the ultimate reason for the order of life processes because the machine itself is maintained in an ordered flow of processes.[7]

To offer a solution to the problems signalized here, von Bertalanffy supplies the model of the open system for the organism.[8] This model avoids the shortcomings of machine thinking without falling, so he believes, into various unscientific speculations about the living organism. The open system model offers a basis for an exact scientific explanation of living phenomena.

With it von Bertalanffy seeks to stake out a position of his own in opposition to the two main currents in biological thought with which he was confronted as a theoretical biologist. In the controversy between mechanists and vitalists, he rejects both of the contending parties. On the one hand, he agrees with the vitalists' criticism that the mechanists fell short in explaining typical phenomena of life. But, on the other hand, he shares the mechanists' skepticism regarding the speculative character of the vitalist notion of entelechies, which elude scientific investigation altogether. An organism is not a machine (contra the mechanists); an organism is also not a machine plus something else ungraspable (contra the vitalists). An organism must be viewed, according to von Bertalanffy, as an open system.

Von Bertalanffy observes that an organism as a whole, resembles a physical-chemical system in equilibrium.[9] In the cell and in the multicellular organism we find a certain composition, a fixed ratio of components that at first sight agrees with the composition of a chemical reaction system in equilibrium. Yet according to von Bertalanffy, while there may be equilibrium systems within it, the organism as such cannot possibly be regarded as such a system. For (thermodynamic) equilibrium is only possible for closed systems. The organism, however, is not a closed system but an open system in a continuous exchange of matter with its environment. "An open system is defined as a system in exchange of matter with its environment, presenting import and export, building-up and breaking-down of its material components."[10]

Considering the living organism as a physical-chemical open system sheds a different light on the fixed ratio of the material components found in the cell or the multicellular organism. Here there is no question of a thermodynamic equilibrium such as is finally attained in a closed system, but rather of a so-called "steady state." Instead of speaking of a "steady state," some writers refer to a dynamic or flowing "equilibrium" (German: *Fliessgleichgewicht*). I use quotation marks here to indicate that in the strict sense there is no equilibrium. Under certain circumstances, an open physical-chemical system can attain such an independent, stationary state, in which the system as a whole remains constant although there is a continuous import and export of matter. Better still than the term "steady state" is the term "quasi-steady state," for the definition of the state of an organism as a steady state applies only as a first approximation, when a shorter period is considered. When the entire life cycle is taken into account, then it is clear that the organism is not in a steady state, but passes through a continuous process of development from the newly formed germ to the dying off of the adult entity.[11]

THE OPEN SYSTEM MODEL MORE CLOSELY CONSIDERED

Criticism like that levelled by von Bertalanffy against machine models for the organism proceeds explicitly or implicitly from a certain conception of what a machine is and how an organism differs fundamentally from a machine. That is true too, of course, about critical evaluations of criticisms of machine models. The entire discussion of the organism as a living machine remains up in the air and ends in confusion if what one is to understand by the term "machine" is left unclear.

We have already seen that in seeking to identify the proper character of the organism, von Bertalanffy asserts the conception of the organism as an open system. In so doing, he thereby also reveals in principle what he sees as characteristic of the machine. For if the organism is an open system and if the concept of the machine is by definition just the opposite, then the machine must be regarded as a closed system. However, the distinction between open and closed systems does not draw a line between living nature and the realm of technology. For while each organism is an open system, or rather consists of a hierarchy of open systems,[12] technology features not only closed systems but open systems as well. It does not escape von Bertalanffy that in chemical technology there are not only batch processes in closed chemical-technical reactors but also open reaction systems with a continuous import and export of materials:

Organisms are open systems. To be sure, they are not the only open systems in existence: a flame is a simple example of a physical system that is "open" (hence the old simile between fire and life); and chemical technology ever more uses open reaction systems in contrast to conventional closed-system or batch processes.[13]

Now, the fact that the open system model is not exclusive for living nature raises the critical question concerning the precise nature of this model and of von Bertalanffy's conception. To be sure, von Bertalanffy would not agree that it is a variant of the *machine* model; but would he not have to accept its classification, more broadly speaking, as a *technological* model, namely, the model of the continuous chemical reactor? Von Bertalanffy will deny that this is so. He argues that the open system model is to be regarded as a specifically biological model and that as such it offers the appropriate basis for an exact explanation of the phenomena of life, such as growth. Open (chemical) systems in nature are much more complex than those in technology. This is so, first of all because in the organism there are reactions between extremely large numbers of components. Moreover, the cell and the organism form, from a physical-chemical standpoint, not homogeneous but extremely heterogeneous colloidal systems. In the organism as an open system there occur, in addition to chemical reactions, a number of other physical-chemical processes such as absorption, diffusion and the like. Hence, the model of the open system is primarily a biological model, but one that in a simplified form is also applicable in technical chemistry.[14]

Von Bertalanffy makes a choice for the conception of the organism as an open system and from there arrives at the definition of a machine as a closed system. The notion that the open system model is a typically biological model, even though it is applicable to certain technical systems, he then defends on the basis of a particular view of the complexity in reality. But why does he insist on regarding as a biological model what others will easily see to be a technological model? His underlying argumentation runs somewhat as follows: Biological systems are more complex than systems of a physical-technical character (machines and other technical constellations). Well then, because the organism is an open system while the closed system is a simplification of the open system (there is no exchange of matter with the environment), physical systems (as a rule) are closed systems. And insofar as open systems occur in (chemical) technology, these are of more modest complexity than the open systems in biology. Clearly, von Bertalanffy proceeds from a particular philosophical understanding of the complexity in reality: he recognizes no *fundamental* boundaries between the biological and the physical-technical "domains" of reality. At the very least, this tends to relativize his fervent opposition to (mechanistic) reductionism.[15]

A second point of criticism concerns the question of the basis of the theoretical conception of the organism as an open system. This conception is inspired by a philosophical view of reality, expressed in the renowned statement of Heraclitus: "You cannot step into the same river twice; for fresh water is forever flowing towards you."[16] This Heraclitean notion touches the fundamental character of life. In von Bertalanffy's view, life must be understood as primarily a stream of life. The forms and structures that manifest themselves in living nature are secondary. There is an essential difference between the forms and structures of inorganic or inanimate nature and those which the biologist studies in living nature. In living nature, the forms as observed from the outside appear to be fixed and enduring. Yet in reality the external shape of the organism is the product of a countless number of processes. Pithily, von Bertalanffy expresses this as follows: "Living forms are not *in being,* they are *happening;* they are the expression of a perpetual stream of matter and energy which passes [through] the organism and at the same time constitutes it."[17]

Von Bertalanffy's basic objection to machine models is that the organism is represented in them as a rigid, static structure similar to what one encounters in inanimate nature. Crystals provide a fine example. While the organism maintains itself in a process of exchange with its environment and is thus subject to constant renewal of the materials of which it is constructed, a crystal consists of the same components throughout the entire duration of its existence, sometimes millions of years.

Traditionally, biology is divided into two parts: morphology and physiology. Morphology is the study of organic forms and structures and physiology is the study of typical processes in living organisms, such as metabolism. A distinguishing characteristic of von Bertalanffy's view of the organism is that it eliminates the usual contrast between structure and function, and between morphology and physiology. Von Bertalanffy rejects this opposition as having a mechanistic origin. Even the structures of the organism (the bearers of the processes) must on his standpoint be conceived as not static but primarily dynamic. Hence the conception of the organism as an open system implies a totally dynamic view of the organism. This applies not only to the organism as a whole but also to its parts, as he makes explicit:

The antithesis between *structure* and *function, morphology* and *physiology,* is based upon a static conception of the organism. In a machine there is a fixed arrangement that can be set in motion, but can also be at rest. In a similar way the pre-established structure of, say, the heart is distinguished from its function, namely, rhythmical contraction. Actually, this separation between a pre-established structure and processes occurring in this structure does not apply to the living organism. For the organism is the expression of an everlasting, orderly process, though, on the other hand, this process is sustained by underlying structures and organized forms. What is described in morphology as organic forms and structures, is in reality a momentary cross-section through a spatial-temporal pattern.[18]

In von Bertalanffy's totally dynamic conception of the organism, the opposition between structure and function is eliminated because both are manifestations of the single flowing stream of life, as he also states explicitly, in the continuation of the previous citation:

What are called structures are slow processes of long duration, functions are quick processes of short duration. If we say that a function such as the contraction of a muscle is performed by a structure, it means that a quick and short process is superimposed on a long lasting and slowly running wave.[19]

MAIN POINTS OF CRITICISM

Both the totally dynamic view of the organism and the open system model of the organism are open to criticism.

In comparing an organism and a machine one can easily observe certain similarities. Both are subject to decay, for instance. However, only an organism can die. The dying off of the organism is an example of an irreversible process which, as such, cannot be compared with shutting down a machine. The difference has to do with the fact that the machine shows a permanence of structure in the absence of its functioning. In an organism, no distinction can be made between a "pre-established structure" and its function. We know already from everyday experience that with the death of an animal, for example, the living structure is lost. What remains after the onset of death is a cadaver that immediately begins to decay, unless action is taken to preserve the remains.

Thus far I agree with von Bertalanffy's dynamic view of the organism. He correctly emphasizes that the organism maintains its structure while it is functioning. My question is whether the totally dynamic view of the organism follows from the unbreakable connection between the structure and function of the organism. I do not think so. I will show in the following that von Bertalanffy's view of the organism is based instead on philosophical assumptions about the fixed and the changeable as these manifest themselves to us in the experience of reality.

With respect to what is fixed, three types may be distinguished. First, there is the fixed character of things, or of what is thing-like in the organism, as opposed to the changeable character of processes and events. Secondly, there is the fixed character of identity: a tree remains itself throughout the duration of its existence, even though undergoing constant change. Finally, there is the fixed character of the functional regularity or order of reality. With respect to what is changeable, two types may be distinguished. Besides the change associated with processes and events as already noted, there is a more comprehensive kind of change that extends to the whole of reality.[20]

A comparison of these distinctions with von Bertalanffy's categories reveals that he acknowledges only the fixed character of the third type, that is, of the functional regularity or order of reality. One must even say that the order of reality is conceived in a restricted sense, namely only with respect to processes. That is clear from his assertion that "what is ultimately persisting is not a lasting structure but a law of a steady process."[21] The consequence of his standpoint must be that the fixed character of the first type, that is, of what is thing-like in the organism (forms and structures), together with the fixed character of the second type, that is, of identity, are "absorbed," as it were, into the changeable character of reality. This is manifest from a statement in which he refers to these two types, although without sharply defining them, in order to affirm that they are not *in being* but rather are *happening:*

> Living forms are not *in being,* they are *happening.*...We believe we remain the same being; in truth hardly anything is left of the material components of our body in a few years; new chemical compounds, new cells and tissues have replaced the present ones.[22]

Von Bertalanffy's dynamic view of the organism rests upon a Heraclitean philosophical perspective. My criticism is that it forces reality into a straitjacket. I am aware that it certainly does not follow from the preceding quotation that von Bertalanffy denies that the "we" still exists year after year in spite of the change in material components. And I agree with him that the identity of the human person cannot reside in the continuing changing material components of our body. But then we still have the problem of how to explain that the human person remains the same through his or her life. Instead of accepting the different types of duration and change as they are given in human experience, von Bertalanffy simplifies reality for the sake of theoretical reconstruction. This results in his choice for the model of the open system. This choice, too, as I shall now explain, is open to criticism.

Von Bertalanffy is aware, as we have seen, that the open system model of the organism also occurs in chemical technology. A continuous chemical-technical reaction process can be described as an open system. Essential for it is the reactor. The reactor does not participate in the reaction process, but it does condition its course. Hence a pure comparison of the organism with the continuous chemical-technical reaction process requires that there be something present in the case of the organism equivalent to the reactor and the control of the process conditions. The question now arises as to what extent an organism may be compared with a reactor since it has a boundary. One may assume that von Bertalanffy's answer would be that the boundary of an organism also consists of living material and is for that reason dynamic. In general, spatial boundaries of systems are indistinct, and perception is not a reliable guide. Von Bertalanffy concludes:

> The spatial boundaries of even what appears to be an obvious object or "thing" actually are indistinct;...the spatial boundaries of a cell or an organism are equally vague because it maintains itself in a flow of molecules entering and leaving and it is difficult to tell just what belongs to the "living system" and what does not. Ultimately all boundaries are dynamic rather than spatial.[23]

So one can say that in comparing the organism with a continuous chemical reaction process, there is no equivalent to be found for the reactor. If, however, one were to take the conditioning reactor into account anyway in applying the open system model to the organism, a new difficulty would appear. The dynamic character of the open system model would be called into question, for the reactor is a fixed, static structure. This leads irrevocably to the conclusion—following von Bertalanffy's own train of thought—that the open system model is also a machine model of the organism.

CONCLUSIONS

The critical reflections in this paper about the open system model for the organism revolve around two fundamental issues. The first issue concerns metaphysical ideas about living phenomena underlying von Bertalanffy's choice of the model of the open system. It has appeared that life (and natural reality in general) is regarded as a dynamic flow of matter and energy. As a result of this metaphysical belief von Bertalanffy argues for a totally dynamic view of the organism eliminating the duality of structure and function. Based on this view, von Bertalanffy launches the open system (the chemical-technical reaction process) as a nonmechanistic model for the organism. The second issue concerns the machine as a metaphor for the organism and especially the quality of the open system model. Our analysis has led to the conclusion that von Bertalanffy's struggle against mechanistic thinking in the sciences has not resulted in finding a way out of its long tradition; the chemical open system model represents, in fact, a new kind of machine metaphor which is inadequate for the different type of open system manifested in organisms. The boundaries of the artificial chemical open system are static, whereas organisms have dynamic boundaries.

NOTES

1. Translated from the Dutch by Herbert Donald Morton.

2. A. Sutter, *Göttliche Maschinen: Die Automaten für Lebendiges bei Descartes, Leibniz, La Mettrie und Kant* (Frankfurt am Main: Athenäum, 1988).

3. G.H.E. Russelman, *Van James Watt tot Sigmund Freud: De Opkomst van het Stuwmodel van de Zelfexpressie* (Deventer: Van Loghum Slaterus, 1983).

4. J.D. Bolter, *Turing's Man: Western Culture in the Computer Age* (Chapel Hill: University of North Carolina Press, 1984).

5. Bolter, 11.

6. S. Strijbos, *Het Technische Wereldbeeld: Een Wijsgerig Onderzoek van het Systeemdenken* (Amsterdam: Buijten and Schipperheijn, 1988).

7. L. von Bertalanffy, *General System Theory: Foundations, Development, Applications* (Harmondsworth: Penguin Books, 1968), 148.

8. In recent decades this model has occupied an important place in discussions of the foundations of biology. Since the mid-seventies the discussion has shifted to the model of the autopoietic system developed by the biologists Maturana and Varela. See H.R. Maturana, and F.J. Varela, *Autopoiesis and Cognition: The Realization of the Living* (Dordrecht, Boston, London: D. Reidel Publishing Company, 1980).

9. L. von Bertalanffy, "Der Organismus als Physikalisches System Betrachtet," *Die Naturwissenschaften* 28 (1940): 521-31. Reprinted as chap. 5 in von Bertalanffy, *General System Theory*.

10. Von Bertalanffy, *General System Theory*, 149.

11. Von Bertalanffy, *General System Theory*, 127, 128.

12. See, for example, von Bertalanffy, *General System Theory*, 168 and L. von Bertalanffy, *Problems of Life: An Evaluation of Modern Biological Thought* (London: Watts and Company, 1952), 129.

13. L. von Bertalanffy, *Robots, Men and Minds: Psychology in the Modern World* (New York: George Braziller, 1967), 73.

14. Von Bertalanffy, *General System Theory*, 130, 149.

15. S. Strijbos, "The Concept of Hierarchy in Contemporary Systems Thinking—A Key to Overcoming Reductionism?" in *Facets of Faith and Science. Volume 3: The Role of Beliefs in the Natural Sciences*, edited by J.M. van der Meer (Lanham: The Pascal Centre for Advanced Studies in Faith and Science/University Press of America, 1996).

16. Von Bertalanffy, *Problems of Life*, 123.

17. Von Bertalanffy, *Problems of Life*, 124.

18. Von Bertalanffy, *Problems of Life*, 134.

19. Von Bertalanffy, *Problems of Life*, 134. Compare von Bertalanffy, *General System Theory*, 25, 172, 261. At these places von Bertalanffy relativizes the contrast between structure and process, also for physical reality (the complementarity of particle and wave).

20. To illustrate the types of fixedness (F1, F2 and F3) and change (C1 and C2) one may compare two entities, namely, a violin and a musical performance. The violin has a thinglike fixed character (F1) in contrast with a musical performance which as an event exists in change during a certain time (C1). At the same time, however, these entities—the violin and the musical performance—both have the fixedness of identity (F2). We speak about this or that violin and this or that musical performance. The sounds produced by the strings of the violin obey certain physical regularities; they are governed by fixed physical laws (F3). Besides the changing character (C1) of certain entities in reality, namely events, one has to notice that also the whole of reality is continuously in change (C2).

21. Von Bertalanffy, *Problems of Life*, 136.

22. Von Bertalanffy, *Problems of Life*, 124.

23. L. von Bertalanffy, *Perspectives on General System Theory: Scientific-Philosophical Studies* (New York: George Braziller, 1975), 165.

11

A Sensible God:
The Bearing of Theology on
Evolutionary Explanation[1]

Paul A. Nelson

> ...paths that a sensible God would never tread...
> S.J. Gould[2]

INTRODUCTION

It is widely held that evolutionary theory, like other natural sciences, partakes necessarily of a methodology according to which one cannot in *scientific* reasoning refer to "God," "the Creator," "creation" (understood as the act of a divine intelligence) or other theological concepts. Evolutionary biologists cite a variety of philosophical doctrines in support of this view, or argue that in all events *methodological naturalism* (as the view has come to be known) stands very much at the foundation of the modern scientific outlook.[3]

Thus it is a fact of considerable interest that, while making a case for evolution—in writing introductory textbooks or encyclopedia articles, for instance—many evolutionary biologists appeal to theology.[4] That is, in presenting a line of evidence or argument for evolution, the author will, as a premise of his argument, *make a theological claim.* The case for evolution, in short, takes an unmistakably theological turn.

In this essay I shall demonstrate that theological arguments affect the content of evolutionary explanations. This is noteworthy on at least two counts. First, the role of theology in evolutionary explanation provides evidence (or counterarguments) against the soundness of the philosophical doctrine of methodological naturalism. Methodological naturalism has lately come under critical scrutiny;[5] this essay provides some raw materials for that philosophical project, by showing how theological premises bear directly on the content of evolutionary explanation.

Second, this essay suggests that the received understanding of such questions as the significance of homologous patterns, or the origin of the genetic code, may be skewed by uncritically accepted theological premises. In justifying evolution, and in explaining the patterns of natural history, biologists have grown accustomed to claims about what a creator would or would not have done.[6] For this practice, they have the example of the *Origin* itself, and indeed, Darwin's writings generally, where arguments of the sort at issue play an important role in the case for evolution.

However, I would encourage biologists to consider Darwin's theological metaphysics with the same careful gaze they have turned on (for instance) his speculations about heredity. To a remarkable extent, Darwin's theological premises or assumptions have become the metaphysical setting for the field of evolutionary biology as a whole, coloring much of what is thought "possible" or "reasonable."

More generally, I urge biologists to reconsider a number of the received arguments for evolution, namely, those resting on unjustified theological assumptions. While the arguments are familiar, their fragility is still largely unappreciated.

In what follows, I begin by examining what I shall call *the argument for evolution from divine wisdom and perfection.* I then turn to what I shall call *the argument for evolution from divine freedom.* After presenting and evaluating each argument, I conclude with a discussion that raises some historical questions.

THE ARGUMENT FROM DIVINE WISDOM AND PERFECTION

In his *History of Creation*, Ernst Haeckel argues that "even if we knew absolutely nothing of the other phenomena of development, we should be obliged to believe in the truth of the Theory of Descent, solely on the ground of the existence of rudimentary organs."[7] Under the heading of "Dysteleology," Haeckel gathered a number of apparently useless or imperfect structures that, he argues, could be reconciled with the theory of creation only by "ludicrous" ad hoc conjectures. In laying great stress on the evidential force of imperfection, Haeckel follows Darwin's lead. Throughout his entire corpus, Darwin is never stronger or more bitter in his language than when condemning the failed teleology of theories of creation, which impute imperfect organic design to the direct intent of a rational and benevolent creator.

Many current evolutionists stand squarely in this tradition. Arguments that trade on intuitions about the nature of God—namely, that from his wisdom and perfection he would create only optimal or perfect designs—occur widely in the recent evolutionary literature, in a variety of contexts.[8] Doubtless the most influential formulations, however, occur in the writings of Stephen Jay Gould. Since many authors draw on Gould's formulations, I will consider them here in detail.[9] While not structured formally as a series of premises leading to a conclusion (Gould is too entertaining a writer to do that), each passage does contain the elements of the argument, and expresses it either implicitly or explicitly. I have emphasized key words and phrases in the passages.

The theory of natural selection would never have replaced the doctrine of divine creation if *evident, admirable design pervaded all organisms.* Charles Darwin understood this, and he focused on features that would be out of place in a world constructed by *perfect wisdom....*Darwin even wrote an entire book on orchids to argue that the structures evolved to ensure fertilization by insects are jerry-built of available parts used by ancestors for other purposes. Orchids are Rube Goldberg machines; *a perfect engineer* would certainly have come up with something better. This principle remains true today. The best illustrations of adaptation by evolution are the ones that strike our intuition as peculiar or bizarre.[10]

Odd arrangements and funny solutions are the proof of evolution—paths that *a sensible God* would never tread but that a natural process, constrained by history, follows perforce.[11]

The proof that evolution, and not the fiat of *a rational agent*, has built organisms lies in the *imperfections* that record a history of descent and refute creation from nothing....Adaptation does not follow the blueprints of *a perfect engineer.*[12]

Evolution lies exposed in the *imperfections* that record a history of descent. Why should a rat run, a bat fly, a porpoise swim, and I type this essay with structures built of the same bones unless we all inherited them from a common ancestor? *An engineer, starting from scratch, could design better limbs in each case.*[13]

But how can a scientist infer history from single objects? This most common of historical dilemmas has a somewhat paradoxical solution. Darwin answers that we must look for *imperfections and oddities*, because any perfection in organic design or ecology obliterates the paths of history and *might have been created as we find it*. This principle of imperfection became Darwin's most common guide....I like to call it the "panda principle" in honor of my favorite example—the highly inefficient, but serviceable, false thumb of the panda.[14]

It will be useful to formalize Gould's argument. I have drawn the following premises from the cited passages.

Premise 1. If *p* is an instance of organic design, then *p* was produced either by a wise creator, or by descent with modification (evolution).

Premise 2. If *p* (an instance of organic design) was produced by a wise creator, then *p* should be perfect (or should exhibit no imperfections).

Premise 3. Organic design *p* is not perfect (or exhibits imperfections).

From these premises, the conclusion follows that therefore organic design *p* was not produced by a wise creator, but by descent with modification. Some organic designs are evidence of evolution.

Note that premises 1 and 2 are theological; they refer directly to a creator, and the actions expected of him. Gould's terms for the creator include "a perfect engineer," "a sensible God," "a rational agent" and "a wise creator." Note further that premises 2 and 3 refer to "perfection," and we may reasonably infer from the cited passages that Gould holds that humans can readily discern the presence or absence of perfection when they examine organic designs.

The conclusion requires of course both that perfection and imperfection be patent qualities of organic design, and that a wise creator would only create perfect organic designs. If these premises are granted, it will follow that any imperfect organic design is not the product of a wise creator. Rather it has come to be via the historically contingent processes of descent with modification.

And, according to Gould, examples of imperfect organic design abound. He writes of "vestigial organs," "odd biogeographic distributions made sensible only as products of history" and "adaptations as contrivances jury-rigged from parts available"[15]—all of which, on the argument from divine wisdom, provide evidence for descent.

Before going on, it is important to note that the "imperfections" at issue are of two types. These are (a) imperfection due to putative descent with modification limited by previous structural constraints, and (b) imperfection due to putative descent with degeneration. A classical example of the former is the pentadactyl limb of tetrapods. An example of the latter might be the vestigial wings of island beetles, or the nonfunctional eyes of certain species of cave fish. Both of course are held to be due to descent with modification. The difference lies in the success of a pattern (homology), or its failure (vestigial organs) to adapt the use of a part to another purpose.

The distinction is useful in that it allows one to focus on precisely which purported aspect of God's character is inconsistent with a given biological pattern (held to be imperfect). The "imperfection" of homologous patterns is often held to be inconsistent with God's *freedom*, for instance, while the "imperfection" of vestigial organs is held to be inconsistent with God's wisdom and perfection. In the sections immediately following, I look at arguments having mainly to do with God's perfection.

SOME PROBLEMS

The argument for evolution from divine wisdom is popular and compelling. It seems to draw on widely shared intuitions about God and the nature and history of the structure of organisms. Discussing the argument with philosophers and biologists, I was struck by how many of them accepted it unreservedly as an impeccable piece of scientific reasoning.

Despite its wide appeal, however, the argument is also deeply problematical. The argument employs theological concepts, such as "a wise creator," and aesthetic or teleological notions, "perfection" and "imperfection," that cannot perform the analytical and empirical work required of them. Each premise of the argument is attended with difficulties.

Premise 1:
"If p is an instance of organic design, then p was produced either by a wise creator, or by descent with modification."

In this section, I will assume that the concept "wise creator" and our ordinary notions of biological perfection and imperfection are unambiguous, that is, they are understood in the same way by all

Facets of Faith and Science

observers (assumptions which will be at issue below). I want first to examine a lesser difficulty, namely, that given our ordinary notions, the first premise of the imperfection argument is a false dichotomy.

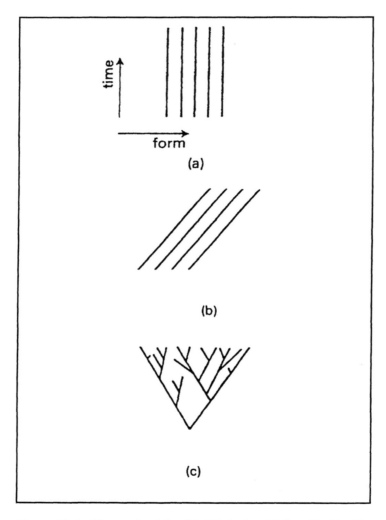

Figure 11.1: Views of origin. Modified after Ridley (1985): *The Problems of Evolution.*

Figure 11.1 (a) depicts what Ridley calls "separate creation," (b) is Lamarckian transformism, and (c) is evolution, or common ancestry. Ridley formulates "separate creation" as stating "that species do not change and that there were as many origins of species as there have been species."[16] Now some creationists may defend this view, although Ridley cites no authority for this interpretation of "species creation." Pattern (a) represents what I will term a *static* theory of creation, in which designs display (more or less exactly) the form in which they were created. One would be hard pressed to find any expression of that view in the creationist literature, whether recently or within the past several decades. Rather, one will find extended discussions of what I will term *dynamic* theories of creation—as represented, for instance, by Figure 11.2.[17] Here, the terminal species

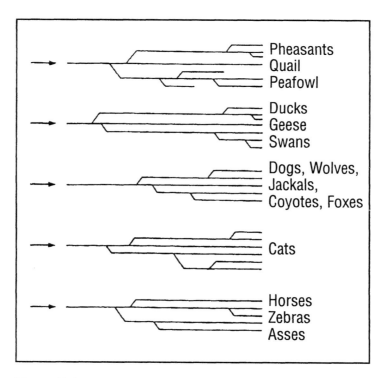

Figure 11.2: Dynamic theory of creation. Modified after Junker and Scherer (1988).

collectively forming a basic type, stem from common ancestors, which were themselves created. Considerable—albeit ultimately bounded—change may have occurred between the creation of a design p and our observation of p. For instance, p may have speciated, or undergone genetic changes which gave rise to phenotype modification. In short, creationists defend the dynamic pattern of Figure 11.2, rather than Ridley's static pattern.

Clearly the argument from divine wisdom presupposes a static theory of creation, akin to 11.1 (a), according to which an organic design p appears today largely as it was originally created. Yet few if any knowledgeable creationists would defend such a theory. In fact, they are likely to acknowledge that some organic designs are biologically imperfect, but will argue that such imperfection is consistent with their theoretical outlook. Not all imperfections, therefore, count against the theory of special creation, or a discontinuous geometry of organic form. In such cases the argument from divine wisdom will fail to hit its mark.

As an example, consider blind cave animals. Evolutionary biologist Douglas Futuyma presents the functionless lens and retina of the cave salamander as instances of the imperfect workings of evolution. He then asks, "Do we find evidence here of wise design?"[18] Yet in the same year that Futuyma posed his question, two well-known creationists, independently considering the same phenomenon, saw it as degenerative change easily understood under a creation theory:

> Blind cave fish with remnants of eyes...appear to have true vestigial organs. These and similar degenerations apparently have indeed resulted from typically disadvantageous mutations...When hereditary changes are small enough to permit survival and reproduction, vestiges may remain. However, these vestigial structures at best are indicative of changes small enough to permit survival and reproduction, vestiges may remain. However, these vestigal structures at best are indicative of changes within limits; they are usually degenerative changes within a species.[19]

So Futuyma's question has answers other than the one he presupposes. In all likelihood the apparently poor design of the cave salamander's eye is the consequence of evolutionary change, but such degenerative changes can readily be accommodated within a dynamic theory of creation.

Consider another example, the rudimentary wings of flightless birds, which Naylor regards as true vestigial structures whose existence contradicts the theory of creation.[20] The Dutch creationist Hendrik Murris, however, in a discussion of genetic drift and the limits of variation, argues:

Suppose that (as an oversimplified example) the allele "A" imparts the ability to fly , while "a" signifies flightlessness. If birds with AA and Aa combinations arrive and breed on an island where they have no natural enemies, the flightless aa individuals which will inevitably be hatched will survive. Some generations later, according to our model experiment, the entire population could be flightless![21]

Murris goes on to argue that within the theory of creation, known genetic processes may explain the origin of some, though not all, species of flightless birds. In a related analysis, the German creationists Reinhard Junker and Siegfried Scherer explain the origin of the rudimentary wings of flightless beetles and insects as cases of degenerative microevolution.[22] In these dynamic theories of creation, extant organic designs are the products not just of original creative intent, but also of the perturbing effects of secondary causes, for example, natural selection, mutation or genetic drift.[23] Thus, in any assessment of the optimality of an organic design, the perturbing effects of secondary and natural causes must be separated from original design (if such a historical reconstruction is possible).

By presupposing a static theory of creation, the first premise of the argument from divine wisdom describes a false dichotomy. Of course, many supposedly imperfect organic designs, such as human upright posture, or the human retina, cannot be explained by a dynamic theory of creation as the consequence of simple degenerative changes. A dynamic theory of creation can accommodate only certain limited neutral or degenerative changes without contradicting another of its main empirical tenets, namely, that genetic and phenotypic variation is bounded. Most "vestigial" structures, for instance, appear to mark out paths of phylogenetic branching that are expressly denied by even the most flexibly dynamic theories of creation. In any event, the argument from divine wisdom need not, indeed should not, presuppose a static theory of creation. That it does so often presuppose such a theory, however, should alert us to the possibility that the argument may rest on other problematical presuppositions.

Premise 2:
"If p *was produced by a wise creator,*
then p *should be perfect. "*

With this premise, we come upon the major theological difficulties of the argument. In Gould's formulations, and in any formulation which includes statements about the character and actions of a "wise creator," the argument from divine wisdom makes theological claims which must be justified or explicated, irrespective of the argument's empirical content. The structure of the argument requires that the referent of "wise creator" be fixed objectively in some way. In other words, any exponent of the argument must explain (1) what a "wise creator" or a "sensible God" is, and (2) what a "wise creator" would do.

To illustrate the first problem, assume for the moment that we are able to identify an imperfect organic design *p*. Then suppose our conception of the creator is similar to John Stuart Mill's: the creator is benevolent and wise but not omnipotent.[24] This creator's power is limited, and thus he would not be able to avoid occasional design compromises. Some imperfections would necessarily be included in the creation—including, let us say, the imperfect organic design *p*. On this view of the creator, the conclusion that imperfection of organic design is evidence of descent with modification would not follow in every case. Gould writes that perfection alone cannot demonstrate descent, because "perfection need not have a history."[25] If we employ Mill's conception of the creator, however, imperfection need not have a history either.[26] If a stapler that continually jams or a water pitcher with a dribbling spout were designed from scratch, they have no history in an evolutionary sense—yet both artifacts are manifestly imperfect to anyone knowing their intended functions. In the absence of any philosophical explication of "wise creator," we cannot exclude Mill's limited creator as a reasonable possibility. There are certainly no empirical grounds on which to do so.

Mill's limited creator is admittedly heterodox (in the Christian tradition), and some may wish to argue that one either defends the usual omnipotent conception of the creator, or one defends no conception at all. The point however is that we have no grounds *within evolutionary theory itself* to exclude Mill's creator, or any one of a number of conceivable creators whose natures allow imperfection. The creator's place in the argument cannot be filled by just any conception.[27] To sustain the conclusion, "Imperfection of organic design is evidence of descent," the argument from divine wisdom requires a particular conception of the creator, namely, the conventional picture of an omnipotent and beneficent artificer (hereafter, the conventional conception). Thus, far from being theologically neutral, the argument has a stake in the truth of a particular theology.

I turn next to the problem of what a "wise creator" would *do*, a problem related to the ambiguity of "perfection" as an operational construct. Suppose we begin with the conventional conception of the creator. According to the second premise of the argument from divine wisdom, if a perfect God created the world, we should expect to observe "perfect" organic design—but what sense should be attached to this term? Is it possible that biological entities judged imperfect when considered individually, might combine to form a macrosystem judged perfect? Here, theological difficulties ordinarily ignored in any biological analysis come crowding to the fore. These difficulties can be avoided, I would argue, only by stipulation.

Consider, for instance, the question of the creator's *proper domain*. Many philosophers and theologians take the creator's proper domain to be the entirety of time and space, and furthermore hold that issues of moral value figure ultimately in any theory of creation. If this is so then the necessary finitude and methodological limits of human scientific observation may lead us to infer mistakenly that an organic design (for example, the panda's pseudothumb) is imperfect, when its imperfection is only apparent, that is, local. On this view, any judgment of perfection or imperfection must be qualified with a proviso that perfection—defined as divinely created perfection—can be judged only on the scale of the whole creation. And there is no reason for a creator to optimize one part of the universe at the expense of the whole. As one commentator writes:

> According to this view, what appears to be evil, when seen in isolation or in a too limited context, is a necessary element in a universe which, viewed as a totality, is wholly good. From the viewpoint of God, who sees timelessly and as a whole the entire moving panorama of created history, the universe is good....[28]

Several philosophers have articulated theodicies which employ just such an analysis; Augustine and Leibniz are notable examples.[29] In his *Theodicy*, Leibniz argues:

> [W]e acknowledge...that God does all the best possible, in accordance with the infinite wisdom which guides his actions....But when we see some broken bone, some piece of animal's flesh, some sprig of a plant, there appears to be nothing but confusion, unless an excellent anatomist observe it: and even he would recognize nothing therein if he had not seen like pieces attached to their whole. It is the same with the government of God: that which we have been able to see hitherto is not a large enough piece for recognition of the beauty and order of the whole.[30]

Although one may regard such a theodicy with skepticism (or scorn: see *Candide*), the problem remains—how is one to judge divine perfection? The question cannot be evaded, for the argument itself demands an answer. To be sure, one can stipulate that only matters of biological optimality are relevant. The stipulation, however, is wholly arbitrary. My intention here should not be mistaken. I am not defending a Leibnizian theodicy. I want only to stress that the "wise creator" of the argument from divine wisdom is hardly the plain and readily employed concept many take it to be. That these problems have been largely ignored by exponents of the argument should not be taken as evidence that they are insignificant.

Premise 3:
"Organic design p is not perfect."

Let me briefly pursue the question of the meaning of "perfection" in a more empirical vein.

All exponents of the argument from divine wisdom hold (either explicitly or implicitly) that perfection and imperfection are observable aspects of organic design. Gould writes of perfection as "the complex and perfected adaptations of organisms to their environments: the butterfly passing for a dead leaf, the bittern for a branch, the superb engineering of a gull aloft or a tuna in the sea."[31] These admirable organic designs are contrasted by Gould with the imperfection of "[r]emnants of the past that don't make sense in present terms—the useless, the odd, the peculiar, the incongruous."[32]

The terms "perfection" and "imperfection" have long been part of the descriptive vocabulary of natural history. We readily apply both to organic designs we admire or find puzzling (or repugnant). As a consequence many authors use the terms with little apparent reflection, perhaps thinking that, as operational constructs in biology, "perfection" and "imperfection" are entirely perspicuous. They are not. The epistemological difficulties that plague optimality arguments in evolutionary theory also occur in judgments of perfection (or imperfection).[33] In the latter case, however, the difficulty of determining whether a state of a trait or organism is optimal is magnified immeasurably by the theological context.

Recall that one premise of the argument (premise 2, in my formalization) states that a "wise creator" will create perfect organic designs. This seems clear enough until we come to particular cases, such as the panda. Gould argues that we can use optimality theory to designate "ideals for assessing natural departures."[34] It follows that in finding existing pandas to be imperfect, Gould must have some notion of an ideal panda, departure from which evokes a judgment of imperfection. So what does an ideal panda look like?

That's rather hard to say, as Maynard Smith has pointed out:

It is clearly impossible to say what is the "best" phenotype unless one knows the range of possibilities. If there were no constraints on what is possible, the best phenotype would live for ever, would be impregnable to predators, would lay eggs at an infinite rate, and so on. It is therefore necessary to specify the set of possible phenotypes, or in some other way describe the limits on what can evolve.[35]

With the argument from divine wisdom, however, the question is not what can possibly evolve, but what can possibly be created—and here a gate opens into an unbounded field of speculation. If we employ the conventional conception of the creator, there seems to be *no* limits on what is possible, nor any reason (short perhaps of logical contradiction) why one hypothetically possible panda should be preferred—as a counterfactual ideal—to another. If "perfection" is limited only by the extent of one's imagination, then specifying an ideal phenotype, for the panda or any other organism, quickly becomes a fanciful exercise. Why could not the creator have given pandas the ability to fly?

We might then turn the problem around. Let us define a criterion of optimality that a "wise creator" ought to be able to achieve and then see whether real organisms measure up. At least that will put some reasonable limits on how we judge perfection of design. Now, however, we must find grounds to explain why, from the set of all possible criteria, we have chosen one particular criterion (or set of criteria). Just as within evolutionary theory, "a proper optimization theory must be capable of explaining why particular constraints on [phenotypic] accessibility are regarded as absolute while others are not,"[36] so the imperfection argument requires some intrinsic reasons why the creator's designs should be limited by physical or biological constraints in certain instances but not in others. This, I would argue, is exceedingly difficult to do. In fact it is impossible.

Take Gould's judgment that the panda's pseudothumb is suboptimal or imperfect, falling short of what we might expect of a "wise creator." Despite this judgment—that the thumb is "somewhat clumsy" and "wins no prizes in an engineer's derby"[37]—Gould writes that while watching pandas at the Washington zoo, he was "amazed at their dexterity, and wondered how the scion of a stock adapted for running could use its hands so adroitly."[38] Indeed, other observers of the panda heap praise on its use of its forelimbs:

The panda can handle bamboo stems *with great precision*, by holding them *as if with forceps* in the hairless groove connecting the pad of the first digit and pseudothumb.[39]

When watching a panda eat leaves...we were always impressed by its *dexterity*. Forepaws and mouth work together with *great precision*, with great economy of motion....[40]

Although the panda's thumb may be suboptimal for many tasks (such as typing), it does seem suited for what appears to be its usual function, stripping bamboo. (At any rate the facts of the matter are very much in dispute.)

But—and here is the nub of the problem—even if the pseudothumb *were* suboptimal for stripping bamboo, *it might still be the best structure possible*. The creator could have been limited in some way by unknown "compossibility" constraints. In crudest outline, a compossibility analysis would ask whether all possibilities are mutually consistent. One cannot, for instance, expect an electric clock designed to obtain its regularity from alternating current to be *more* regular than that current.[41] Or the thumb may have some unknown primary function for which it was designed, and the panda has co-opted it secondarily to strip bamboo. One may have failed to identify the correct reference situation by which to judge the design (perhaps by looking at too narrow a slice of the panda's life history). The flippers of marine turtles, for example, strike us as rather badly designed for digging holes in beach sand to place eggs. The same flippers, however, perform efficiently in the water, where the turtles spend most of their time. Which reference situation takes precedence in an optimality analysis?[42]

If we allow that the creator need only "act reasonably," that is, create organic designs which meet some specific criteria for optimality, then we must be able to say what those criteria are, and why they obtain, if our claims of suboptimality or imperfection are to have any evidential force. This problem is made acute by the bothersome truth that any suboptimal design can be made optimal if we add the right constraints.[43] What principles, then, guide us in specifying reasonable criteria of optimality for an omnipotent creator?

A simple equation illustrates the problem. Suppose we define an optimal organism (design) as scoring 1.0, where the observed and expected design values in the following equation correspond exactly:

$$\frac{\text{observed design}}{\text{expected design}} = \text{optimality (suboptimality) measure}$$

Now suboptimality enters in when the numerator value falls below that of the denominator. Thus, if an optimal (created or ideal) panda has an expected design value of, say, 50, but actual pandas score 30, the panda as a species is suboptimal, suffering what we might call a "design shortfall":

$$\frac{\text{observed design} - 30}{\text{expected design} - 50} = 0.6 \text{ design shortfall}$$

We cannot solve this equation, however, without the expected design value. Absent the denominator, the equation has two unknowns. One equation with two unknowns is unsolvable.[44] But the expected design must be determined by optimality criteria, a set of metrics along which design is measured—and we have no such metrics for living things as divinely created. Thus we have no principled way of assigning the expected design value.

Evolutionists have learned to be wary of facile arguments about optimality and perfection within evolutionary theory. The divergence of views on the panda's pseudothumb, given above, is a good example of why they are wary. Gould finds the structure "somewhat clumsy," whereas G. Schaller, H. Jinchu, P. Wenshi and Z. Jing give it the precision of a "forceps." Would these investigators differ so widely on the question of, say, the panda's diploid karyotype number?

Many evolutionists continue nonetheless to employ "perfection" and "imperfection" in arguments for evolution. Gould repeatedly uses the word "proof" for the imperfection argument. That, it surely is not. Presumably, arguments for evolution are not constructed out of whatever rhetorically useful notions happen to be available, whether those notions are intelligible or not. Presumably, if "perfection" is a problematical notion within evolutionary theory, it will be a no less problematical notion when employed in an argument in support of the theory.

THE ARGUMENT FROM DIVINE FREEDOM

On opening any moderately advanced biology textbook nowadays one is likely to find, amid the discussion of the evidence for common descent, an illustration showing an array of tetrapod forelimbs (see Figure 11.3). The accompanying text will state that the pattern of similarity abstracted from a comparison of the limbs (the pentadactyl limb) can be explained only by common descent—a scientific advance credited to Darwin.

Francisco Ayala, for instance, in his *Encyclopedia Britannica* article on evolution, writes:

> From a purely practical point of view, it is incomprehensible that a turtle should swim, a horse run, a person write, and a bird or bat fly with structures built of the same bones. An engineer could design better limbs

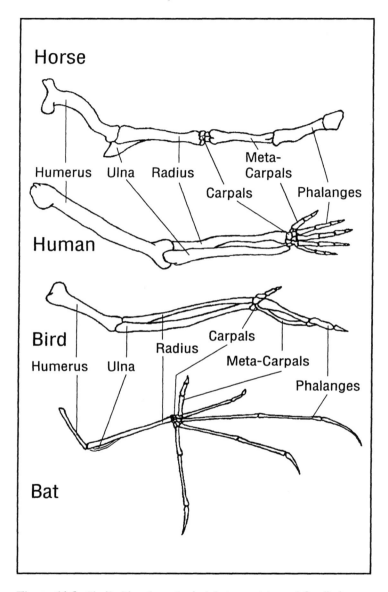

Figure 11.3: Similarities (homologies) between tetrapod forelimbs.

in each case. But if it is accepted that all of these skeletons inherited their structures from a common ancestor and became modified only as they adapted to different ways of life, the similarity of their structures makes sense.[45]

In Chapter XIII of the *Origin*, Darwin argued that it would be "hopeless" to explain the pentadactyl pattern "by utility or by the doctrine of final causes."[46] Darwin's view of these patterns is now canonical within evolution: "Darwin...originated the evolutionary interpretation which has been followed ever since, that the general plan of the pentadactyl limb is not now adaptive, although it must have been in the common ancestor, but its modifications are adaptive."[47] Gould expresses the point as Darwin did, by reference to the forelimbs of mammals, adding (somewhat more bluntly than Darwin did) that the forelimbs fall short of what we might expect from an optimizing designer: "Why should a rat run, a bat fly, a porpoise swim, and I type this essay with structures built of the same bones unless we all inherited them from a common ancestor? An engineer, starting from scratch, could design better limbs in each case."[48]

Notice that, expressed this way, the argument is a variant of the argument from divine wisdom. The force of the argument is increased by the apparent continuity of the imperfect structures across taxa.

This perception of suboptimality is widely shared among evolutionary biologists. However the quality at stake is often not so much imperfection as *lack of imagination*. Here again, I would argue, strong theological preconceptions are at work, coupled with a view of the character of scientific explanation. If we suppose that the creator is free to do as he pleases, the appearance of *plan* can readily become the appearance of *limitation* or *constraint*, suggesting an unimaginative or even slavish repetition of structures along the lines of some predetermined pattern. "Intelligence and purpose," writes Neal Gillespie, interpreting Darwin's arguments against creation, "should be more creative than nature showed itself to be."[49] Moreover, explanations of patterns of homology in terms of a divine plan would violate the rule that explanations in science should be in terms of natural causes, not supernatural ones.

Consider an example from molecular biology. It is widely held that all organisms descended from a common ancestor because they share certain biochemical "universals," such as the genetic code.[50] These biochemical or molecular universals are generally regarded as a "strong evolutionary prediction."[51] As Futuyma puts it,

The only possible reason for these chemical universalities is that living things got stuck with the first system that worked for them. Once the genetic code was established, no species was ever free to try a new one. A mutation that caused the nucleotide sequence UUU to code for glycine instead of phenylalanine would have messed up all the species' proteins.[52]

On the other hand (argues evolutionary biologist Mark Ridley, developing the same point as Futuyma), if "different species had all been *created* separately, we should be very surprised if they had all been built with exactly the same genetic code."[53] We should be very surprised—to supply the missing, but implied, premise—because a *freely acting* creator could have constructed any number of different codes: "Where a Creator would have been free to use different biochemical building blocks, evolution was not free: the history of the earliest organisms determined everything that happened thereafter."[54] This biochemical variant of the argument holds that the apparent uniformity of certain biological patterns, such as the genetic code, is inconsistent with the freedom of a divine creator to act as he wishes. Therefore those patterns evolved.

The theological premise in this argument—that the apparent uniformity of certain biological patterns is inconsistent with the freedom of a creator to act as he wishes—is nowhere better illustrated than in Darwin's book on the "contrivances" of orchids. After reviewing the homologies of orchids and ordinary flowers, Darwin appeals to our intuitions about *what God would have done in this case*:

> Can we feel satisfied by saying that each Orchid was created, exactly as we now see it, on a certain "ideal type:" that the omnipotent Creator, having fixed on one plan for the whole Order, did not depart from this plan: that he, therefore, made the same organ to perform diverse functions—often of trifling importance compared with their proper function—converted other organs into mere purposeless rudiments, and arranged all as if they had to stand separate, and then made them cohere? Is it not a more simple and intelligible view that all the Orchideae owe what they have in common, to descent from some monocotyledonous plant...?[55]

Removing the theology from Darwin's argument for the common descent of the Orchideae would eviscerate it. Darwin provides no fossil evidence that orchids evolved from ordinary flowers, nor indeed any experimental evidence that such a transformation is even possible.[56] Rather, in the chapter leading to the passage cited above, Darwin describes patterns of similarity among orchids—which patterns might, on a creationist reading of the evidence, indicate the purposeful workings of a designer. If one accepts however the premise that it is *unfitting* to ascribe variations on an "ideal type" to the direct artifice of an omnipotent creator, the same patterns become evidence of common descent. The theology in the passage is thus far more than a rhetorical device. It is the logical pivot of Darwin's entire argument.

Arguing in this vein, Futuyma points out that existing animals could have been designed otherwise than they are: "An omnipotent creator could, as we can in imagination, create organisms with wings on their shoulders (for example, angels, or the mythical Pegasus), but the wings of all flying vertebrates are modifications of the front legs of their ancestors."[57] *How uncreative to have done things that way* is the principal deliverance of this species of biological imagination. To be told that the evidence suggests nevertheless that the structures in question were created (that is, are transformationally discrete, not standing in an historical pathway of material descent) does not of course answer the question, "but why should they then appear to share the same plan?" It is this question after all that many biologists want answered. One can imagine that organisms *could* have been constructed in any number of ways.

I find it odd that speculations about the freedom of the Creator should ground an empirical argument for evolution. Surely the reason that homology is regarded as evidence for descent is not because it contradicts what one would expect a rational creator to do. It is true that an omnipotent creator would have been free to make nonhomologous vertebrate limbs, or to give organisms different genetic codes or indeed to make organisms out of different types of matter entirely. These are hardly grounds however to support an empirical claim about the causal history of homologous patterns, and the taxa that exhibit those patterns. Suppose one argues, contra Darwin, that we have every reason for thinking a creator would have designed each species of orchid to show homologies with ordinary flowers. How, by everyday scientific methods, would one go about settling this dispute? One may assume or deny the truth of Darwin's particular theological aesthetic, but it is hard to see how that assumption is binding on other observers (or why we should take it as intelligible in any event).

Many formulations of the homology argument for descent, however, rest on similar theological presuppositions. It is curious that in glossing the orchid arguments, Gould and other commentators (for example, Michael Ghiselin)[58] have not noticed this problem. Perhaps Darwin's theological aesthetic is so closely congruent with their own intuitions that its role in the argument escapes comment. Nevertheless, these theological assumptions need to be justified, or else it should be admitted that they stand as bare postulates. In analyzing Darwin's argument about homologies—in particular, his claim that a creator would not use such patterns—Løvtrup observes: "Why not? Even the Creator may use a good device more than once. Yes, why not indeed?

Darwin's arguments against this possibility are postulates, unfounded by any evidence."[59]

If the phenomenon of homology provides evidence for descent with modification, it must do so *not* because homologies are inconsistent with what a rational creator would have done—a rational creator might have done any number of things—but rather because homologies appear to mark out a pathway of natural transformation that characterizes a continuous geometry of organic form (that is, of material descent). The relevant empirical question is whether the appearance of natural transformation is more than an appearance: have such modifications of biological form occurred? Is the geometry of nature profoundly continuous? These questions, to be somewhat naive about it, want empirical answers. Speculations about the freedom of the creator need to be seen for what they are, and moved out of the discussion.

DISCUSSION:
THE INFLUENCE OF DARWINIAN METAPHYSICS

Darwin's argument for descent with modification was pressed on many fronts—among them, the theological. As several historical and philosophical analyses of Darwin's corpus have noted,[60] the *Notebooks*, the *Sketches* of 1842 and 1844, and the *Origin* itself are permeated by a metaphysical program which was, Cornell argues, "more than useful rhetoric to Darwin, and more than a methodological convention that promoted science."[61]

Consider one important aspect of that program, the notion of "perfection." Dov Ospovat writes:

> The assumption of perfect adaptation, which Darwin shared with most of the biologists of his generation, was derived from the belief that nature is a created, harmonious, and purposeful whole...This is the assumption...that organisms are as well fitted as possible for the conditions under which they live. This assumption, in one form or another, was held by virtually every naturalist and natural theologian of the mid-nineteenth century. It is a natural, perhaps necessary, corollary of the belief that nature is a harmonious system preplanned in every detail by a wise and benevolent God.[62]

Cornell concurs:

> The word "perfect" is an adjective generally reserved for divine action. That is how, for instance, Paley used it, and it was probably what Darwin understood, even when he was criticizing the belief in the

perfection of particular forms...because that belief implied special creation by God.[63]

Now, while Darwin came to reject the idea that organisms were perfectly designed for their environments, *he never rejected the theoretical apparatus implied by the very terms "perfection" and "imperfection."* The numerous theological arguments in the *Origin* make sense only if one adopts certain premises, about the nature of the creator, from the classical argument from design. Darwin does not challenge the orthodox (or, in my term, conventional) conception of the creator, defended by his creationist opponents. Rather, he turns to certain aspects of organic design which appear to fit only awkwardly into the usual schemes of natural theology, and drives these organic counterexamples back into the machinery of the argument from design. Instead of impiously attacking the nature or existence of the creator (as a skeptic like Hume, might do), Darwin offers his theory of descent (and secondary causes) to explain what would otherwise—on the conventional view of creation—be intolerable anomalies. *All this incongruity of design could not have been directly created.*

In so doing, of course, Darwin impales his creationist opponents on the horns of a dilemma. Either they deny the benevolence and wisdom of the creator, by making him the author of "abhorrent" designs, or they retain their wise and benevolent conception of the creator, but must greatly circumscribe his actions, for if imperfect designs could be due to secondary causes, then could not many other (in fact, nearly all) organic structures be the products of secondary causes as well?

But note again that in all this there is little to indicate that Darwin ever rejected the deep presuppositions which he inherited from English natural theology, namely, perfection as an observable quality of organic design, and the orthodox or conventional conception of the nature (if not the actions) of the creator. Indeed, a close reading of the *Notebooks* would seem to suggest that Darwin saw his theory as providing a *more* sublime conception of the actions of the creator (see, for instance, D 36: "What a magnificent view one can take of the world..."). Many passages in Darwin's corpus make little sense unless it is acknowledged that Darwin was employing *a particular conception of the creator to judge the theories of his creative activity.* Otherwise, why should the multiple creations scornfully derided in D 36 as a "long succession of vile molluscous animals" be beneath the "dignity" of God? Cornell argues, of this and other passages from later notebooks:

As always, Darwin's idea of "perfection" refers to the nice relationship of organisms to their physical surroundings. But it also refers to the overall design of the world, from a divine viewpoint...Darwin's sense of a comprehensive system, the invocation of divine perfection, and his new theory are thus all closely related.[64]

Brooke argues, in relation to these passages:

The fact is that there are several entries in the transmutation notebooks which indicate that Darwin was discovering a philosophy of nature which he genuinely believed conferred a new grandeur on the deity, despite—or rather because of—the fact that it superseded Paley.[65]

While current evolutionists may not share (or in fact may be opposed to) Darwin's theological motivations, their use of the imperfection and homology arguments for descent presupposes the intelligibility of notions rooted in Darwin's theological metaphysics. The notion of perfection as an observable quality of organic design, and the intuition lying at the heart of Darwin's metaphysics—that a rational and benevolent God would have created an organic world different from the one we observe—continue to inform the philosophical foundations of evolutionary theory (as should be evident from the passages I have cited from Gould above).[66]

Yet it is widely held that an important aspect of the Darwinian revolution was the surrender by biologists and natural historians of any epistemic warrant to theological speculation in science.[67] Indeed, many scientists and philosophers would argue that natural science and theology view each other across a largely (if not completely) impassable epistemological gulf.[68] Science, on this view, is by its very nature committed to a thoroughgoing methodological naturalism. Hence, the problem which opened this essay: the persistence of Darwinian theological themata in current evolutionary theory is prima facie inconsistent with the doctrine of methodological naturalism.

But should natural science necessarily be committed to methodological naturalism? One might argue that the shortcomings of theological arguments for evolution are evidence enough that science has no business meddling in theology (or vice versa). I take a different moral away from these questions, however. *Science will have to deal with theological problems if science is a truth-seeking enterprise; theology must confront the patterns of scientific experience if it hopes to speak to all of reality.* What this essay helps to show, I think, is how very easy it will be to do both theology and science badly. That is not a brief for methodological naturalism, however. It is a tale of caution about how we should go about explaining the origin of the world's creatures.

NOTES

1. I thank Bill Wimsatt, Bob Richards, Leigh Van Valen, Douglas Allchin, Bob Schadewald, Michael Ruse, Stephen Gould, Kurt Wise, John Cornell, Dan McShea, Marc Swetlitz, Chris Cosans, Steve Meyer, Bill Dembski, Phil Johnson, Roy Brown, Gabriel Nelson, Jitse van der Meer and students and faculty in the Department of Philosophy, and the Committee on the Conceptual Foundations of Science at the University of Chicago, for helpful discussions; and Frank Arduini, Tom Bethell, Ron Brady, Mike Cavanaugh, Kevin Padian, Elliott Sober, Bill Hasker, and four anonymous referees for helpful written comments. I benefitted from reading an unpublished essay by Sarvar Patel. Earlier versions of this paper were presented at the Adventist Forum of Atlantic Union College, the Second International Conference on Creationism, the Wheaton College Second Annual Theological Conference, Kentucky State University, the First International Pascal Centre Conference on Science and Belief, and to a faculty seminar at Biola University. All the usual disclaimers apply about any of the above persons or places being responsible for the content of the paper.

2. S.J. Gould, *The Panda's Thumb* (New York: W.W. Norton, 1980).

3. See, e.g., F.J. Ayala and J. Valentine, *Evolving: The Theory and Processes of Organic Evolution* (Menlo Park: Benjamin/Cummings, 1979), 5; N. Eldredge and J. Cracraft, *Phylogenetic Patterns and the Evolutionary Process* (New York: Columbia University Press, 1980), 3; S. Beck, "Natural Science and Creationist Theology," *BioScience* 32 (1982), 739-40; D. Futuyma, *Science on Trial* (New York: Pantheon Books, 1983), 169-70; A. Riddiford and D. Penny, "The Scientific Status of Modern Evolutionary Theory," in *Evolutionary Theory: Paths into the Future*, edited by J.W. Pollard (New York: John Wiley, 1984), 18; S.J. Gould, "Darwinism Defined: The Difference between Fact and Theory," *Discover* (January 1987): 70; A. Hoffman, *Arguments on Evolution* (Oxford: Oxford University Press, 1989), 11-12.

4. By "evolution" I mean what Doolittle calls the "widely accepted" theory that "all Life on Earth today is descended from a common ancestral organism that existed sometime between 1.5 and 3.0 billion years ago"— this evolution occurring by means of the natural selection of undirected and randomly arising variation (R.R. Doolittle, "New Perspectives on Evolution Provided by Protein Sequences," in *New Perspectives on Evolution*, edited by L. Warren and H. Koprowski [New York: Alan Liss, 1991], 165.)

5. From P.E. Johnson, *Reason in the Balance: The Case Against Naturalism in Science* (Downer's Grove: InterVarsity Press, 1995); A. Plantinga, "When Faith and Reason Clash: Evolution and the Bible," *Christian Scholar's Review* 21 (1991): 8-32 and "Methodological Naturalism," in *Facets of Faith and Science. Volume 1: Historiography and Modes of Interaction*, edited by J.M. van der Meer (Lanham: The Pascal Centre for Advanced Studies in Faith

and Science/University Press of America, 1996); J.M. van der Meer, "Beliefs in Science: Taking the Measure of Methodological Materialism," *Proceedings of the Second Annual Wheaton Theology Conference* (Wheaton: Wheaton College Press, in press) and "The Struggle between Christian Theism, Metaphysical Naturalism and Relativism: How to Proceed in Science?" *Faculty Dialogue* 26 (in print); J.P. Moreland, "Theistic Science and Methodological Naturalism," in *The Creation Hypothesis: Scientific Evidence for an Intelligent Designer*, edited by J.P. Moreland (Downers Grove: InterVarsity Press, 1994), 41-66 among others.

6. In the spring of 1979, I attended a series of lectures on evolution at the Carnegie Museum of Natural History, in Pittsburgh. The first lecture in the series was given by Leonard Krishtalka, a vertebrate paleontologist on the museum staff. For his opening illustration, Krishtalka had borrowed a peccary (a pig-like mammal) from the museum's collection, which he placed on the dais at the front of the lecture hall. Pointing to the peccary's "dew claws" (so-called because these toes, on the rear of the limb just above the hoof, appear to touch only the "dewy" surface of the ground), Krishtalka asked, "Now why would God have created this animal with nonfunctional structures like the dew claw?" But of course God did not create the peccary, he continued, natural selection did. What strikes me now about this illustration was how utterly clear its theological content seemed to Krishtalka, that is, as evidence supporting the causal story he was about to tell.

7. E. Haeckel, *The History of Creation* (New York: D. Appleton, 1876), 291.

8. See, e.g., F. Jacob, "Evolution as Tinkering," in *The Possible and the Actual* (New York: Pantheon Books, 1982); E. Sober, *The Nature of Selection* (Cambridge: MIT Press, 1984), 175-76; D. Futuyma, "Evolution as Fact and Theory," *Bios* 56 (1985): 6; R. Dawkins, *The Blind Watchmaker: Why the Evidence of Evolution Reveals a Universe without Design* (New York: W.W. Norton, 1986), 91-94; R. Burian, "Why the Panda Provides no Comfort to the Creationist," *Philosophica* 37 (1986): 11-25.

9. Some readers of the early drafts of this paper argued that I ought not to cite Gould's general and semipopular essays, on the grounds that the essays were general and semipopular. The more I thought about this argument, however, the odder it seemed. The argument means either (a) Gould has knowingly misrepresented evolutionary theory to his lay readers, or (b) in presenting evolutionary theory to his lay readers, Gould has used language so "analogous, rhetorical, sometimes imprecise" (to quote one of my correspondents) that he has—despite his best intentions—misrepresented evolutionary theory. Those who make this argument can hardly have meant (a), but is (b) any more credible? On the topic of biological imperfection, Gould's popular (and technical) writings are strikingly consistent. It is hard to believe that a scientist renowned for his prose abilities would explain his own theory so poorly that he could, for more than a decade, continually mislead his

readers. And Gould takes his popular writing far more seriously than many of his colleagues. "The concepts of science, in all their richness and ambiguity," he argues, "can be presented without any compromise, without any simplification counting as distortion, in language accessible to all intelligent people" (*Wonderful Life* [New York: W.W. Norton, 1989], 16). It is instructive to note that the reading list for Gould's introductory science course at Harvard ("History of Life," Science B-16) includes, out of a total of 62 readings, 33 of his essays from *Natural History* magazine.

10. S.J. Gould, *Ever Since Darwin* (New York: W.W. Norton, 1977), 91.

11. Gould, *Panda's Thumb*, 20-21.

12. S.J. Gould, *Hen's Teeth and Horse's Toes* (New York: W.W. Norton, 1983), 160, 164.

13. Gould, *Hen's Teeth and Horse's Toes*, 258.

14. S.J. Gould, "Evolution and the Triumph of Homology, or Why History Matters," *American Scientist* 74 (1986): 63.

15. Gould, "Triumph of Homology," 64.

16. M. Ridley, *The Problems of Evolution* (Oxford: Oxford University Press, 1985), 3.

17. R. Junker and S. Scherer, *Entstehung und Geschichte der Lebewesen* (Giessen: Weyel, 1988).

18. D. Futuyma, *Science on Trial* (New York: Pantheon Books, 1983), 198.

19. W. Frair and P. Davis, *A Case for Creation* (Chicago: Moody Press, 1983), 29.

20. B. Naylor, "Vestigial Organs are Evidence of Evolution," *Evolutionary Theory* 6 (1982): 94.

21. H.R. Murris, "Concept of the Species and its Formation" in Concepts of Creationism, edited by E.H. Andrews, W. Gitt, and W.J. Ouweneel (Welwyn Herts: Evangelical Press, 1986), 200-1.

22. Junker and Scherer, *Geschichte der Lebewesen*, 126. Junker later authored a systematic treatment of rudimentary organs and atavisms within a creationist framework. R. Junker, *Rudimentäre Organe und Atavismen: Konstruktionsfehler des Lebens?* (Berlin: Studium Integrale/Zeitjournal Verlag, 1989).

23. C. Darnborough, "Genes—Created but Evolving," *Concepts in Creationism*, edited by E.H. Andrews, W. Gitt, and W.J. Ouweneel (Welwyn Herts: Evangelical Press, 1986), 252-62.

24. See J.S. Mill's *Essays on Ethics, Religion and Society*, edited by J.M. Robson (Toronto: University of Toronto Press, 1969) in particular, the essay on "Theism," Part I, especially the introduction and chapter 6, and Part II, "Attributes." Bertrand Russell discusses a similar conception of the creator: "He need not be omnipotent or omniscient; He may be only vastly wiser and more powerful than we are. The evils in the world may be due to his limited power. Some theologians have made use of these possibilities in forming their conception of God" (B. Russell, *History of Western Philosophy* [New York: Simon and Schuster, 1945]; 589).

25. Gould, "Panda's Thumb," 28.

26. Thanks to Frank Arduini for this phrase.

27. Hume's *Dialogues Concerning Natural Religion*, especially Part V, is the classical expression of the difficulties that follow from considering the full range of logically possible supreme beings (creators).

28. J. Hick, "Evil, The Problem of," in *The Encyclopedia of Philosophy*, edited by Paul Edwards (New York: Macmillan Publishers, 1967), 137.

29. See Augustine, *Confessions*, Book VII, section 13 (New York: Penquin Classics, 1961), 148-49.

30. G.W. Leibniz, *Theodicy* (1710; La Salle: Open Court Press, 1985), 206-207.

31. Gould, *Panda's Thumb*, 28, emphasis added.

32. Gould, *Panda's Thumb*, 28-29.

33. The literature on optimality theory and its difficulties is extensive. The "epistemological difficulties" I have in mind are best explained by Richard Lewontin, in his essay "The Shape of Optimality," in *The Latest on the Best*, edited by J. Dupre (Cambridge: MIT Press, 1987), 151-59.

34. Gould, "Triumph of Homology," 66.

35. J.M. Smith, "Optimization Theory in Evolution," *Annual Review of Ecology and Systematics* 9 (1978): 32. Under the heading, "What is possible?" R.M. Alexander takes up the same question:

> The next problem, after deciding what is likely to be optimized, is to decide what structures or strategies are possible, and what constraints apply. If no such limitations were recognized it would have to be concluded that optimum structure would make bones unbreakable and without mass, and an optimum life-history would involve immortality and infinite fecundity.

Optima for Animals [London: Edward Arnold, 1982], 97.

36. Lewontin, "The Shape of Optimality," 156.

37. Gould, *Panda's Thumb*, 24.

38. Gould, *Panda's Thumb*, 21.

39. G. Schaller, H. Jinchu, P. Wenshi and Z. Jing, *The Giant Pandas of Wolong* (Chicago: University of Chicago Press, 1986), 4; emphasis added.

40. Schaller, Jinchu, Wenshi and Jing, 58; emphasis added.

41. See N. Rescher, *Leibniz: An Introduction to his Philosophy* (Oxford: Basil Blackwell, 1979), chap. 6, for an explication of such compossibility arguments within Leibniz's philosophy.

42. R.C. Lewontin, "Adaptation," in *Conceptual Issues in Evolutionary Biology*, edited by E. Sober (Cambridge: MIT Press, 1984), 234-51.

43. Lewontin, "Shape of Optimality," 158-59.

44. Gabriel Nelson kindly pointed this out to me.

45. F.J. Ayala, "Evolution, The Theory of," *Encyclopedia Britannica*, 15th ed. (Chicago: Encyclopedia Britannica, 1988), 987, emphasis added.

46. Darwin, *Origin of Species*, 435.

47. A.J. Cain, "The Perfection of Animals," in *Viewpoints in Biology*, vol. 3, edited by J.D. Carthy and C.L. Duddington (London: Butterworths 1964), 44.

48. S.J. Gould, "Evolution as Fact and Theory," *Discover* 2 (1981): 36.

49. N. Gillespie, *Charles Darwin and the Problem of Creation* (Chicago: University of Chicago Press, 1979), 71.

50. The genetic code is the 64 trinucleotide code (often represented in a table showing a U [uracil] C [cysteine] A [adenine] G [guanine] nucleotide matrix) by which messenger and transfer RNA specify the sequence of amino acids in protein assembly. Thus, UUU and UUC code for the amino acid phenylalanine, AAA and AAG code for lysine, and so on. For examples of the argument for common descent from the genetic code, see A.G. Cairns-Smith, *The Life Puzzle* (Edinburgh: Oliver and Boyd, 1971) 148; T. Dobzhansky, "Nothing in Biology Makes Sense Except in the Light of Evolution," *American Biology Teacher* 35 (1973): 127; J.M. Smith, *The Theory of Evolution* (New York: Penguin, 1975), 82; T. Dobzhansky, F. Ayala, G.L. Stebbins and J. Valentine, *Evolution* (San Francisco: W.H. Freeman, 1977), 28; D. Futuyma, *Evolutionary Biology* (Sunderland: Sinauer, 1979), 38; E. Mayr, "Darwin, Intellectual Revolutionary," in *Evolution from Molecules to Men*, edited by D.S. Bendall (Cambridge: Cambridge University Press, 1983), 30-31; D. Raup and J. Valentine, "Multiple Origins of Life," *Proceedings of the National Academy of Sciences USA* 80 (1983): 2981; B. Davis, "Molecular Genetics and the Foundations of Evolution," *Perspectives in Biology and*

Medicine 28 (1985): 256; Dawkins, *The Blind Watchmaker* (1986), 270;
M. Ridley, *Evolution and Classification* (London: Longman, 1986), 119-20;
C. Patterson, "The Impact of Evolutionary Theory on Systematics," in
Prospects in Systematics, edited by D.L. Hawksworth (Oxford: Clarendon
Press, 1988), 61; E. Sober, *Reconstructing the Past* (Cambridge: MIT Press,
1988), 9; A. Hoffman, *Arguments on Evolution* (Oxford: Oxford University
Press, 1989), 8-9; E. Mayr, *One Long Argument* (Cambridge: Harvard
University Press, 1991), 23.

51. B. Davis, "Molecular Genetics and the Foundations of Evolution,"
Perspectives in Biology and Medicine 28 (1985): 256.

52. D. Futuyma, *Science on Trial* (New York: Pantheon Books, 1983), 205.
Since 1985 several variant nuclear codes have been discovered, leading many
workers to doubt the theory, employed here by Futuyma, that once established
the code must be invariant. See T.D. Fox, "Natural Variation in the Genetic
Code," *Annual Review of Ecology and Systematics* 21 (1987): 67-91, F. Caron,
"Eucaryotic Codes," *Experientia* 46 (1990): 1106-17; and T.H. Jukes and
S. Osawa, "Recent Evidence for Evolution of the Genetic Code," in *Evolution
of Life*, edited by S. Osawa and T. Honjo (New York: Springer-Verlag, 1991),
79-95.

53. M. Ridley, *The Problems of Evolution* (Oxford: Oxford University
Press, 1985), 10; emphasis added.

54. Futuyma, *Science on Trial*, 205.

55. C. Darwin, *On the Various Contrivances by Which Orchids Are
Fertilized by Insects* (1877; Chicago: University of Chicago Press, 1984), 245-
46.

56. Darwin performed numerous experiments (the description of which
constitutes much of the book) to understand the functional aspects of orchid
fertilization. In the last chapter, he draws this moral from the experiments:

> No one who has not studied Orchids would have suspected that these and
> very many other small details of structure were of the highest importance
> to each species; and that consequently, if the species were exposed to
> new conditions of life, and the structure of the several parts varied ever
> so little, the smallest details of structure might readily be acquired
> through natural selection. *These cases afford a good lesson of caution
> with respect to the importance of apparently trifling particulars of
> structure in other organic beings.*

Various Contrivances, 287; emphasis added.

57. D. Futuyma, "Evolution as Fact and Theory," *Bios* 56 (1985): 6.

58. See M. Ghiselin, *The Triumph of the Darwinian Method*, 2nd ed.
(Chicago: University of Chicago Press, 1984).

59. S. Løvtrup, *Darwinism, The Refutation of a Myth* (London: Croom Helm, 1987), 132.

60. See, e.g., Gillespie, *Charles Darwin and the Problem of Creation*; J.H. Brooke, "The Relations Between Darwin's Science and his Religion," in *Darwinism and Divinity*, edited by J. Durant (London: Basil Blackwell, 1985), 40-75; and Cornell, "God's Magnificent Law."

61. Cornell, "God's Magnificent Law," 384-85.

62. D. Ospovat, "God and Natural Selection: The Darwinian Idea of Design," *Journal of the History of Biology* 13 (1980): 189-90.

63. Cornell, "God's Magnificent Law," 396.

64. Cornell, "God's Magnificent Law," 397.

65. Brooke, "Relation between Darwin's Science and his Religion," 46.

66. Some readers of this essay have urged me to note that Gould, like many (most?) evolutionary theorists, is himself an agnostic, or, as Gould has put it, a "nontheist." In a symposium dealing in part with this topic, Jeffrey Levinton observed that Gould "is about as 'theological' as anyone in this room calling himself an atheist." All true, no doubt, and all beside the point. What Gould or any other advocate of the imperfection argument believes in his heart of hearts is immaterial. The theological convictions of a person have no bearing on the truth or falsity of the propositions, theological or otherwise, that he puts forward publicly as worthy of assent.

67. Mayr, "Darwin, Intellectual Revolutionary," 25.

68. The literature on this question is vast; for a representative selection of opinion, see L. Kolakowski, *Religion* (Oxford: Oxford University Press, 1982); L. Gilkey, *Creationism on Trial* (Minneapolis: Winston Press, 1985); K. Yandell, "Protestant Theology and Natural Science in the Twentieth Century," in *God and Nature*, edited by D.C. Lindberg and R.L. Numbers (Berkeley: University of California Press, 1986), 448-71; and S.J. Gould, "Darwinism Defined," 70.

12

Mind and Brain, Science and Religion: Belief and Neuroscience in Donald M. MacKay and Roger W. Sperry[1]

Marvin J. McDonald

This essay highlights the contributions of Donald M. MacKay and Roger W. Sperry to understanding the interrelations of belief and science in the twentieth century. These two prominent neuroscientists spent considerable energy exploring relationships between religion, broadly conceived, and the conduct of science. In fact, their high profiles as scientists provided them with more than usual opportunity to reflect upon and explore in print their views on science, religion and values. Each scholar also sought to interpret the significance of twentieth-century neuroscience for the communities outside of their scientific specialties. Because of these factors, this joint case study can rely more upon published materials than would be possible in the lives of most scientists.[2]

The following two sections of this essay outline MacKay's and Sperry's theory of mind-brain relations, their philosophies of science, and their models of science-belief relations. In the concluding section I highlight several features of the relationships between belief and science evident in the work of these two scholars.[3]

DONALD MACKAY:
SCIENTIST, CHRISTIAN, PHILOSOPHER

MacKay's approach, consistently developed throughout his career, was grounded in information theory.[4] Information theory provided the linchpin for his entire framework, from his philosophy of science to his research program to his view of Christianity-science relations.

MacKay's Theory of Consciousness[5]

For MacKay, the brain scientist studying consciousness confronts two fundamentally different kinds of "facts": the "inside story"—the personal experiences of an agent, and the "outside story"—the observations of a detached (that is, "objective") scientist on brains and their functions. Each set of data is accessible from two different perspectives, that of the agent and of the observer, respectively. The task of a theory of consciousness is to provide an adequate account of the correspondences between these two sets of data while acknowledging their dependence upon the inside and outside perspectives. Many attempts at a theory of consciousness have been hampered, in MacKay's view, by confusions about the categorical distinction between these two kinds of facts,[6] the nature of human beings, and the nature of science. MacKay's theory of consciousness was offered as a contribution to solving the first two problems.

The importance of understanding the categorical distinction between mental and brain activities arises from some unique complexities of human nature. MacKay's key anthropological claim is that a person is so complex ontologically that the perspectives of the agent and observer by themselves each miss some real aspects of the conscious agent.[7] In particular, the phenomena of conscious agency cannot be directly detected by an outside observer, and an agent is not aware of his or her own brain functioning. Furthermore, each perspective is constrained in such a way as to prevent the simple concatenation of observer and agent perspectives into a monolithic view of the conscious agent.[8] This is true even though the aspects accessible to the inside and outside perspectives are all real features of the same unitary entity, a conscious agent.

MacKay's solution to these complexities is to use information theory as a conceptual "bridge" between the two categories of events, mental and neural. He started with the neuroscientist's working assumption of a correlation between neural and mental events.[9] Information theory then allows a theorist to explore the correlation while maintaining the distinction. This strategy provides the background for his theory of conscious agency.

MacKay begins his theory of consciousness by defining agency as "activity under evaluation, activity in view of an end or purpose or goal."[10] To this definition MacKay added another criterion to differentiate conscious agency from the unconscious agency of simple pupil reflexes, for example. Conscious agency requires "self-supervisory activity" as well. Two functions are implied by this phrase, and both are activities engaged in by a system vis-à-vis itself: (a) the setting of behavioral priorities (which are internal to the system) in an ongoing manner, and (b) the maintenance of "conditional readinesses" for action and the planning of action. The latter function is seen as a "matching response to the demands of sensory input" in perceptual processes.[11] The hypothesis of self-supervisory activity is offered, then, as a necessary feature of any neurological system proposed as the correlate of experience since these properties constitute minimal features of conscious agency. This criterion is employed as a guide to ongoing research and is intended to provide some necessary but insufficient conditions to be satisfied by any account of conscious experience.

A second key thesis in MacKay's theory of consciousness is the mutual determination of neural processes and mental activities. The familiar manner in which brain activity determines mental activity is uniformly accepted by neuroscientists, including MacKay.[12] The recognition of the "causal efficacy" of mental activities is less widespread, so mutual determination distinguishes MacKay's approach from many current views in the brain sciences.

In MacKay's view the principles of self-supervisory activity and mutual determination clarify the nature of the correlation between mental and neural events while maintaining the inside/outside distinction and his key anthropological claims. Information theory "bridges" mental and neural events by identifying self-supervisory activity as a set of properties which is characteristic of both sets of phenomena. The principle of mutual determination further elaborates the nature of that correlation by specifying the bidirectionality of influences between neural and mental processes.

MacKay was aware of the controversial nature of his claims. Many critics attempted to collapse the distinction between mental and neural events. He consistently responds to such attempts in two ways. First, he points out that in his research the distinction between inside data (yielded by reports of conscious experience) and outside data (arising from observation of the brain) was not redundant.[13] Moreover, his theory of consciousness demonstrates a manner in which neuroscience can maintain the distinction while exploring the correlation. Therefore confusing mental and neural events is scientifically unwise. Second, MacKay argues that attempts to conflate the mental and the neural arise from confusions about the relationship between scientific observation and personal experience. These converging arguments show that MacKay's theory of consciousness and his philosophy of science have significant implications for one another.

MacKay's Philosophy of Science[14]

Although the inside story and the outside story both yield data important for brain science, MacKay clearly states that the epistemological foundation of science is provided by detached observation. For MacKay, this situation raises questions about the nature of the relationship between scientific observation and the perspective of the agent. The key concept in MacKay's response to these questions is the logical relativity of the standpoints of the agent and the observer.[15]

MacKay often relies upon thought experiments to establish this notion. Imagine a neurophysiologist who has exhaustive knowledge of the brain as a completely physically determined system.[16] The question MacKay raises is, "Does the knowledge of the scientist have an unequivocal claim upon the assent of the agent being studied?" In other words, MacKay is interested in the relationship between the knowledge claims of an observer with access to a fully deterministic brain science and the knowledge (warranted beliefs) an agent has of his or her own experience. Is the agent logically compelled to accept the prediction of the brain scientist about his or her own future behavior?[17]

MacKay's negative answer to this question depends upon two fundamental steps in the argument: (a) once the brain scientist communicates with the agent by presenting the prediction, then there are physical changes in the brain of the agent associated with the very act of communication and with the agent's subsequent belief or disbelief of the prediction;[18] (b) the prediction is invalidated because another influence (the communication of the prediction to the agent), not taken into account in the original prediction formulated by the observer, has changed the system to which the prediction applies (the agent and his or her brain).

MacKay also considers a next step in this thought experiment often brought up in critiques of logical relativity. What happens if the scientific observer modifies the deterministic prediction to take into account what happens when the agent understands and believes the scientist's prediction? In this instance, the modified prediction is correct for both the scientist and the agent *if and only if* the agent believes the modified prediction. That is, the agent is logically correct to believe the modified prediction. But MacKay also claims that the agent would be logically correct to disbelieve the modified prediction because it is predicated upon the assumption of the agent's belief. He concludes, "There does not exist one and only one specification of the immediate future of your cognitive mechanism with an unconditional claim to your assent."[19]

A crucial consequence of this argument for MacKay is that the future specification of an agent's cognitive mechanism is open and undetermined by other neurological processes (for the agent) so that any future specification is "a future that *waits to be determined* by what [the agent] thinks or decides."[20] Neither of these points necessarily hold true for the scientific observer *per se*, however counterintuitive it might seem.[21]

The future events determined by the agent include experience, behavior and neural functioning, since they are all aspects of the conscious agent. In particular, the agent's determination of the future involves determination of neural activity by mental activity. So logical relativity supports a key thesis in MacKay's theory of consciousness, the mental determination of brain processes. Outside the specific context of predictions about an agent's cognitive mechanism, the processes of neural determination of mental events remains comprehensible, thus maintaining the thesis of mutual determination. Logical relativity also has significant implications for MacKay's understanding of causality.

In his theory of consciousness, the neural level is approached in physiological theory, an informational level is characterized by information theory, and a mental level is accounted for in the categories of personal experience. MacKay describes the nature of cross-level determination displayed in logical relativity as analogous to information flow. The parallel he frequently uses to explain mental determination of brain activities is the process by which the form of a mathematical equation in a program shapes or determines the electronic behavior of the computer in which the equation is embodied. So the form of mental activities has consequences for the shape of neural activities.[22] Information flows "cause" other information flows within the same level, but saying that mental events "cause" neural events invites the confusion of intra and interlevel relations, so he uses the more general term "determine" for all relationships of influence. "Cause" could then be reserved for intralevel relationships and "interdependence" could describe the interlevel relationship between mental and neurological processes.[23] So the distinction between interdependence and causality enables MacKay to emphasize that the correlation of mental and neural events involves an interlevel relationship.

The logical relativity of the standpoints of the observer and the agent shows the fallacy of attempting to collapse the ontological distinction between the mental and the neural. Any such attempt leads to contradiction. This follows from the different warrants which are true for the agent and the observer regarding the observer's prediction of the future of the agent's cognitive mechanism. If one simply conjoins the perspectives of the agent and observer, the differences in truth status of the prediction for the two perspectives' status yields a contradiction. This argument converges with MacKay's effort to maintain the distinction between the mental and the neural in his empirical and theoretical work.

To summarize, MacKay's theory of consciousness and philosophy of science reinforce one another. On the one hand, the analogy provided by information theory elucidates the way in which an agent can determine future events. On the other hand, logical relativity implies the mutual determination of mental and neural events and helps identify the correlation of the mental and neural as an interlevel relationship (interdependence). Finally, logical relativity also sustains the ontological complexity of the conscious agent in MacKay's anthropology.

MacKay found that some readers attempted to interpret logical relativity as implying a radical disjunction between the mental and the neural. Such a reading contradicts the correlation between and mutual determination of mental and neural events as well as MacKay's anthropology. He consistently responds to such interpretations by emphasizing the correlation of mental and neurological processes. Two concepts in his philosophy of science are employed in this context: embodiment and complementarity.

The embodiment of mental events in neural processes is a one-to-many correlation between mental and neural processes in a conscious agent. Embodiment also includes the relationship between the software and hardware of a computer. In a conscious agent and in a computer, the upper level processes (mental activities, software) are said to be embodied in the lower level phenomena (brain processes and hardware, respectively). MacKay holds "embodiment" as a tentative thesis.[24]

The notion of complementarity is the epistemological counterpart to embodiment.[25] Complementarity is the relationship between two true but apparently "irreconcilable" descriptions of an entity: for example, the physicists' and consumers' standpoints vis-à-vis an electronic sign advertising hamburgers. Each complementary standpoint is exhaustive in that all details of an entity are accounted for, but each is incomplete in the sense that some questions appropriate to the entity are unanswerable from within a given perspective.

The notions of embodiment and complementarity allow MacKay to state succinctly the task of a theory of consciousness: neuroscience explores the embodiment of mental events in the brain from the perspective of the scientific observer in a fashion complementary to the standpoint of a conscious agent.[26] This formulation counteracts interpretations of logical relativity as compartmentalization.[27] Logical relativity does not impair the capacity of brain scientists working within an outside perspective to capitalize on data from both mental and neural events. It does, however, point out that scientific understanding is not exhaustive of all knowledge. Neuroscientific knowledge is complementary to the understanding an agent has of his or her own agency. This outcome does not reflect a failure of science. Instead, it is a consequence of the ontological complexity of conscious agents.

In conclusion, MacKay's philosophy of science converges with his theory of consciousness in the mutual determination of mental and neural events, the value of information theory and the ontological complexity of the conscious agent. Moreover, logical relativity demonstrates the inherently nonexhaustive nature of science as a human enterprise.[28] Neuroscience employs the standpoint of the observer but cannot entirely subsume the perspective of the agent without generating contradictions or other confusions.[29] Yet the perspectives of the agent and the observer are complementary, not compartmentalized, since mental events are embodied in the brain.

MacKay on Christianity-Science Relations

Two features of MacKay's views on the interaction between science and Christianity are important for this paper: (a) constraints on their interaction, and (b) forms available for fruitful interaction.[30] Concerning constraints upon science-Christianity relations MacKay writes:

> There are no facts of brain science known to me that add up to any kind of "proof" of the existence of God, nor would I expect such a conclusion to follow logically from data of that kind. Science is by its own constitution agnostic. On the other hand there is a real need in our day to bring back a sense of awe into our attitude to the natural world. The invitation is to marvel at the existence of the whole world, the bits whose interconnections we see as well as those we don't understand. This kind of wonder, which no amount of scientific understanding can remove, is quite distinct from and not to be confused with the scientific puzzlement

of the exploring scientist which it is his whole ambition to reduce. Nevertheless, there are not a few people in our day, some of them in high academic places, who seem to think that our slowly growing understanding of the mechanisms of the human brain makes it more difficult to take seriously the religious claims that speculative theology would consider to be at least open, and that revealed theology, especially in the Judeo-Christian tradition, would affirm to be sober truth.[31]

MacKay's view of science and theology displays both his philosophical notions and his theory of consciousness. The standpoint of the detached observer in science is complementary to that of the religious believer as a conscious agent. Christian theology assumes the stance of the religious believer and so provides a complementary perspective to that of the scientist.[32] Therefore, science is agnostic with respect to many claims made by theologies of the "Judeo-Christian" traditions. This agnosticism prevents ontological reductionism (denying the reality of *religious* experience) and the associated category errors.[33] So logical relativity limits the ontological import of scientific knowledge for theology and grounds the compatibility of science and Christianity. For example, concerning life after death, MacKay believes that the embodiment of human consciousness in the brain is logically consistent with a notion of disembodied existence after death, even though theologically he finds reembodied existence to be more plausible.[34] Neuroscience cannot reasonably pontificate on the nature of resurrected humans.

In addition to defending a believer's standpoint against people usurping the name of science in support of anti-Christian views, MacKay also wants to protect science from misguided directives by Christian believers. He repeatedly asserts the ideal of scientific objectivity as detached observation, for example.[35] And his versions of interdependence, embodiment and complementarity leave methodological determinism, methodological reductionism and mechanistic theories intact and at the heart of scientific method. In these and other ways, MacKay's philosophy of science and his version of Christianity-science relations reinforces the scientific mores of the "rigorous" traditions he prefers.

MacKay shows that science not only comports well with Christianity, but that a mutually enhancing, tentative dialogue is also possible between them. For example, MacKay speculates that spiritual activities by human beings are embodied in psychological processes in a manner parallel to the embodiment of psychological processes in neurological activities. Spiritual activities might "show up" in psychological correlates as changes in the patterns of a person's values and priorities, and perhaps in the way a person lives out those priorities. The relation between spiritual and conscious activity which occurs in dialogue with God could involve the emergence of a new "level" of spiritual activity identifiable by a feature of information flow among the "forms" of psychological activities.[36]

ROGER SPERRY:
SCIENTIST AND ADVOCATE FOR WORLD UNITY[37]

Sperry's Theory of Consciousness

Sperry's theory of consciousness relies upon the key notions of "emergence" and "downward causation." In developing his notion of emergence, Sperry frequently uses the emergence of molecular properties from the interactions of constituent atoms to illustrate the point. Sperry clearly characterizes the "spatiotemporal relations" or patterns among parts as the ontological basis of emergent properties, including mental properties, and insists repeatedly upon the causal significance of these patterns and upon the reality of these mental properties.[38] Sperry admits, however, that this model of emergence may make feasible in principle a fully reductive account of the emergent phenomenon of pain in terms of the spatiotemporal arrangement of nerve impulses, but that it remains fantastically impractical.[39]

Sperry clearly asserts his belief in the causal impact of mental phenomena, calling it "downward causation."[40] Sperry relies upon the notion of "neural circuit" to explain his notions of causality. A neural circuit is a dynamic pattern of neural activity that coheres functionally in relation to the overall pattern of neural activity. Some neural functions are performed by large networks of neurons acting in a coherent pattern, and those functions collectively instantiate a control hierarchy in the brain. Sperry hypothesizes that mental phenomena simply *are* the functioning of higher order neural circuits. In the same sense that a molecule is a "new" emergent entity emerging from lawful patterns of atomic interaction, likewise a mental phenomenon is an emergent (functional) entity constituted by the activities of higher order neural circuits.

Downward causation is the influence of the emergent properties upon lower level components (like neurons) and their activities. Thus mental phenomena are completely embedded in the causal nexus of physical reality;[41] they are not nonphysical processes that insert themselves into gaps in the causal nexus of spatiotemporal reality (as dualists would have it). And mental phenomena are real, not epiphenomenal, and not *merely* a neurological component of the brain (as in identity theory) since they are located in "higher" (supraneuronal) levels of the control hierarchy of the brain. Sperry is careful to highlight the "two-way" causation implied by his model.[42] The physiological mechanisms continue to function without interruption, and some causal patterns from physiological levels of activity shape mental phenomena even while the same physiological processes are "simply carried along or shunted this way and that" by mental processes.[43] And the mental and neurological levels are just two from among a full hierarchy of levels.[44]

In short, Sperry's model of consciousness can generally be seen as a kind of functionalism.[45] That is, emergent properties are seen as constituted by the coherent behavior of lower-level components. Both Sperry's and MacKay's models of consciousness emphasize the mutual determination of mental and neural processes. And like MacKay, Sperry sees his version of "downward causation" as having strong implications for our understanding of the nature of science. These authors tend to emphasize differences between their views, although they never reach a consensus on the exact nature of the relationship between their theories.[46]

Sperry's Philosophy of Science

Sperry, like MacKay, claims that his theory of consciousness occupies a "middle road" between materialist identity theses and dualistic interactionist models. Beyond his debates with philosophers,[47] Sperry developed his philosophy of science (often referred to by Sperry as a "metatheory" or "paradigm") in reaction to psychological behaviorism and in agreement with the "cognitive revolution."[48]

Sperry offers his emergentist theory of consciousness as a supplement to reductionism, materialism, determinism, atomism, mechanism, antisubjectivism, antirealism of values and the noncausal status of mental phenomena.[49] He emphasizes that this expansion does not compromise the standards of objectivity he sees as fundamental to science.[50] Furthermore, Sperry sees the implications of his emergentism as reaching far beyond the brain sciences and psychology to all sciences and to the philosophy of science.[51]

For instance, Sperry believes that in an emergentist understanding of mental activities the free will/determinism debate simply vanishes as a pseudoproblem. "We still inhabit a deterministic universe, but it is ruled by a large array of different types, qualities, and levels of determinism."[52] That is, volition is fully determined, but the determining causes issue from multiple levels, not just from neurological levels as reductionists would have it. Sperry emphasizes multiple levels of determination of free will for two basic reasons. First, he wants to highlight the causal efficacy of subjective phenomena such as values via the processes of downward causation within the causal nexus of the cosmos. Second, Sperry wants to emphasize the sensitivity of the brain and mental processes to multiple causes, a principle which satisfies the concerns of voluntarists.[53] Sperry claims that he supplements the traditional versions of upward causation ("microdeterminism" of the reductive materialist variety) with the emergent determination of downward causality. So Sperry's solution arises by wedding traditional microdeterminism with a macrodeterminist version of free will. Overall, the ramifications of his theory of consciousness which have most clearly caught his imagination are the values implications of this model of consciousness.[54] It is in this context that Sperry develops his scientism most clearly.

Sperry on Science-Religion Relationships

For Sperry religion has to do with values. "Liberal religion" obtains its values from an evolutionary worldview generated by science. Liberal religion is defined as religion that does not rely on dualistic or supernatural beliefs.[55] Supernatural religious traditions are rejected because the values fostered by them are responsible for the negative influences of religion in history and in the world today: lack of respect for the biosphere, holy wars, overpopulation, intolerance and so forth.[56] He claims that

> The value differences that derive from these various belief systems...exert profound and pervasive effects on social and political decision-making that add up to what is probably one of the most, if not *the* most formidable of the divisive influences that now confront us and operate against world harmony and unity.[57]

In these assertions religion is an ultimate frame of reference which encompasses a value system involving all of life.

Sperry's version of science preempts religion since values and an entire worldview can be derived from science. Sperry argues for

> a unifying ultimate frame of reference for social values...founded in the revelations of science. In other words, founded [sic] in a reference framework based on empirical evidence and the scientific method as the best avenue to truth available to the human brain and also on the world view that derives from science, i.e., the world model and view of reality

supported by the total collective knowledge of all the sciences along with the insights and perspectives this knowledge brings.[58]

This pathway from neuroscience to religion is paved by the notion of downward causation. In Sperry's emergentist theory of consciousness, values and beliefs causally impact behavior and thus all of society. Since science can explain downward causation, mental events and all human activity can be explained by science. Moreover, scientific knowledge provides the "best avenue to truth" so scientific explanations give us the best possible picture of reality. Religion, seen as values, is therefore fully subsumed by science. Liberal religion consists in acknowledging this ontological and epistemological preeminence of science.

Sperry's preemption of religion by science clearly illustrates his scientism. Science offers "a cosmic scheme and view of life"[59] which will provide the common core of all viable religions in a unified world.[60] Moreover, Sperry places scientific knowledge at the heart of all human life, not just religion. The survival of the human race in the face of current challenges rests on the possibility of the world uniting around values derived from scientific knowledge.[61] Scientific theory provides the foundation upon which all human knowledge and activities can safely rest, so all of life can be approached within one vast epistemological framework or worldview. Downward causation provides the basis for bridging the traditional dichotomies between science and the humanities, science and religion. Science, the humanities and religion are all transformed by the merger, but all of the major traditions of each are incorporated into this "grand synthesis" unchanged. In contrast, MacKay's logical relativity emphasized a fundamental plurality of standpoints.

INTERRELATIONS OF BELIEF AND SCIENCE IN MACKAY AND SPERRY

To identify specific patterns of belief-science coherence in Sperry's and MacKay's views, I first describe the fundamental identity of their theories of consciousness. Next I place that theory in the context of their divergent philosophies of science and views on determinism. Then I examine a specific instance where, in the course of brief published exchanges, these scientists drew contrasting metaphysical implications from the same theoretical principles. In conclusion I list a few of the questions raised by this preliminary exploration of belief-science interrelations in the work of MacKay and Sperry.

Consciousness and the Brain in MacKay and Sperry

MacKay's and Sperry's emergentist theories of consciousness both incorporate the mutual determination of mental and neural processes.[62] The key assertion in this notion is that mental events are real and can have downwardly causal effects.[63] Sperry and MacKay both agree that these determinations involve cross-level relationships. Downward causation is the impact of spatiotemporal patterns of an emergent entity as a unit (for example, a rolling wheel) upon the components of the entity (for example, by shaping the trajectories of its constituent atoms). This description by Sperry converges with the analogy to informational causality offered by MacKay. Mental events form or shape neural events as part of the process of controlling behavior. And they agree that when neural events influence the higher-level processes in the brain (as in brain damage or intoxication), then neural events influence mental events. Moreover, they agree that both the inside data from conscious experience and the outside data from the observations of brains are indispensable to neuroscience. But Sperry holds that science subsumes the perspective of the conscious agent while MacKay insists that the agent's standpoint is complementary to that of the scientist as observer.[64]

This theory of consciousness rules out both reductionist ontologies and the classical "substance" dualism of Cartesian traditions. That is, Sperry and MacKay assert that mental processes are ontologically emergent and downwardly causal (contra reductionists) but that they are not a nonphysical "substance" of some kind (contra dualists). For MacKay the ontological distinction between mental and neural events prevents the reduction of the mental to the neural. This distinction was supported by the empirical fruitfulness of perceptual illusions in his research and by the logical relativity of the standpoints of the agent and observer.[65] Logical relativity also sustained the downward causation thesis, based on the mental determination of neural activity, another contrast with the reductionist move. Contrary to the dualist claims about the nonphysical nature of the mental, MacKay holds to the embodiment of the mental in the physical realm and to the capacity of information theory to bridge the two realms from within an observer's perspective. Likewise Sperry's emphasis on the emergence of the mental in the activity of neural circuits asserts irreducibility and downward causation (nonreduction) while his functionalism counters the dualistic thesis. Thus they both claim to have identified a "middle road" between the alternatives of reductive materialism and substance dualism.[66]

Despite their agreement on concepts characterizing the brain-mind relation, however, they have very different understandings of science.

According to MacKay the "inside" perspective of the conscious agent is never directly accessible to the observer whose knowledge is, therefore, limited. For instance, the realms of values or of religious experience can yield inside data (for example, person X believes Y is wrong), but because these data must be viewed from the outside perspective science must be agnostic about certain truth claims within those realms (is Y wrong?). Sperry's scientism, on the other hand, asserts that the standpoint of the conscious agent is subsumed when neuroscience can specify precisely which functions of higher order neural circuits constitute which mental activities.[67]

Moreover, differences emerged between Sperry and MacKay about the nature of free will and choice despite their commitment to the same substantive theory of consciousness. MacKay asserts that a scientist with an exhaustive and fully deterministic theory of brain function could never, in principle, fully account for all choices and beliefs of a conscious agent. The inherent logical relativity of the observer's and agent's perspectives is the key to identifying this limitation of metaphysical determinism. For Sperry, on the other hand, the emergence of mental phenomena over neural events cannot imply that choice somehow transcends the explanatory framework of science. Volition is causally determined in the future grand synthesis of the sciences which fully incorporates the principles of downward causation and emergence. This is so because the assumption of a closed causal nexus is fundamental to his view of science, and so it follows that future elaborations of neuroscientific theory will be fully congruent with this version of causal determinism. Sperry emphasizes, however, that the variety of causally determining events comports well with voluntarism. Desires and beliefs (which are also functional patterns of coherence in brain circuits) are major determiners of our choices, as one would expect. And neural events per se come into the account primarily as one expands the scope of explanation to include larger portions of the causal nexus of the world.

For Sperry and MacKay, then, the same fundamental principles in a neuroscientific theory of mind-brain relationships coexist with differing metaphysical beliefs about determinism and free will and with divergent philosophies of science. This is so despite their agreement upon the requirements of methodological determinism, methodological reductionism and of detached observation as the norm for objectivity, for example. These patterns of difference and similarity raise many questions about the nature of science-belief interaction. How can the same theories of consciousness be held by scholars with such divergent views on choice and the nature of science? Are metaphysical beliefs about choice dependent upon the religious commitments of the scholar or vice versa? How strongly is the philosophy of science related to the model of science-religion relations developed by each writer? Which of these views shape and influence which others? The purpose of this paper is to show the promise of studying the patterns of science-belief interaction in these two scholars. In a preliminary analysis, these questions cannot be answered. Nonetheless, these questions invite further consideration and demonstrate the richness of this case study for displaying science-belief dynamics. In particular, the striking patterns of divergence between a scientific theory and metaphysical beliefs beg for investigation.

To illustrate possible directions for inquiry, consider first the question of the place of MacKay's anthropology in his overall outlook. One can explore the role of MacKay's anthropology by working out the implications of a modified version of his anthropology for the other elements of his system. For instance, consider the implications if one assumes that the complexity of the conscious agent can be characterized as the operation of multiple levels of functioning organized into a functional hierarchy: molecular activity, cellular activity, organ-level activity, individual organism-level activity and so forth. For each level of activity, one can define information flows and energy transfers, so MacKay's theory of consciousness becomes embedded in a larger theory of biological information systems. Is this move congruent with the rest of his system or not? MacKay's principles of complementarity and embodiment apply to interlevel relations besides the mental-neural one, but logical relativity does not extend in the same way. Even these brief comments can provide a glimpse into a direction which would reward further study. After conducting further analysis, we would be better able to identify the patterns of connection among MacKay's key concepts and assumptions.

Other important directions for inquiry raised by this case study encompass broader questions in science-belief interaction. For instance, can the coexistence of a shared theory of mind-brain relations with different metaphysical beliefs and philosophies of science be explained as a case of different interpretations of the same phenomenology of mind-brain interaction?[68] If so, what broader contexts can account for the difference between the interpretations? At least three important contexts invite investigation. First, many scientists claim that their research experiences are crucial determinants of such interpretive differences. Would this rationale remain plausible for MacKay and Sperry in light of their respective research programs?[69] Second, the social, intellectual and institutional contexts of European and North American neuroscience communities diverge in ways which might be significant in this case.[70] Finally, which broader philosophical contexts for the views of these two scholars could have shaped their interpretive tendencies?[71] These empirical, social and philosophical contexts all show promising directions for further analysis of this case.

Moreover, the exposition of Sperry and MacKay presented above is embryonic. Additional details from the case underscore further disjunctions between their scientific theory and their metaphysical beliefs. A final example delineates connections among their religious beliefs, their models of science-religion relationships and their theory of consciousness.

Metaphysical Implications of Embodied Consciousness: MacKay versus Sperry

In 1980 Sperry and MacKay began a public exchange of views.[72] This exchange flowed in part from discussions they had during MacKay's visit to Sperry's lab in 1979-80.[73] The content of the disagreement and the rationales they provided to support their views are both revealing. This last example highlights possible roles of religious beliefs and models of science-religion relations in the overall pattern of science-belief relations.

The question at issue in the exchange between Sperry and MacKay was whether the embodiment of mental activities in brains is consistent with a belief in consciousness during life after death.[74] In MacKay's view, a scientific understanding of the embodiment of consciousness in brains implies nothing specific about the possibility of conscious life after death. Sperry disagrees, claiming that embodiment contradicts the possibility of human consciousness surviving death. Their disagreement persists despite the fact that they share a similar understanding of embodiment. That is, they drew opposite implications from downward causation. MacKay sees no conflict between downward causation and a belief in human life after death.[75] Yet Sperry describes his position on the emergent downward causation of consciousness as a monism that "restricts its answers to one-world dimensions and says 'no' to an independent existence of conscious mind apart from the functioning brain."[76]

An analysis of science-belief interactions in this exchange raises a multitude of questions. One natural starting point revolves around the significance of personal religious beliefs in this debate. MacKay envisions the possibility of a person being recreated after death, complete with physical embodiment, but Sperry entertains no such option. And these beliefs correspond clearly to the different implications each scholar draws from the theory of consciousness.

But how might one investigate the pattern and extent of the influence of religious beliefs in detail? An important slant on this question emerges by examining the rationales offered by Sperry and MacKay for their conclusions. MacKay asserts the compatibility of a religious belief in an afterlife with his theory of consciousness. His argument is straightforward. The embodiment of mental activities in neural activities means that functioning brains are a necessary condition of the existence of a conscious agent, yet as a scientific thesis, embodiment pertains only to the nature of persons in the domain of space-time encountered by science. Extrapolating the notion of embodiment to a domain of life after death depends solely on the way one's metaphysical beliefs and philosophy of science define the relationship of the afterlife with the domain of space-time as characterized by science. Therefore many disparate views of life after death are consistent with embodiment given appropriate metaphysical beliefs and corresponding assumptions about the nature of science. MacKay's theory of consciousness cannot directly refute or support a belief in the afterlife, given his view of science as a delimited, nonexhaustive enterprise that does not necessarily extrapolate beyond the current cosmos.[77]

Sperry, on the other hand, maintains that the inseparability of consciousness and brain function disallows afterlife notions. This contention arises from his scientism which makes scientific knowledge normative for all human knowledge. So causal determinism, and any other crucial scientific principle, become a standard by which all values and beliefs are to be judged.[78] It is also possible, in Sperry's model, for the theory of consciousness to legitimate or support a religious doctrine if it coincides with values sustained by science, for instance. These rationales display a strong interdependence among the authors' models of science-religion relations, their philosophies of science and the implications which are drawn from scientific theory.

But how might these rationales reflect influences from religious beliefs? Neither account acknowledges such influences directly. Yet the brief exposition of Sperry and MacKay suggests that a potentially rich interplay exists among religious beliefs, research experience, methodological heuristics, philosophies of science and metaphysical beliefs. To illustrate one direction in which an analysis of the likely roles of these various factors might proceed, consider how MacKay's religious beliefs might relate to his arguments in the exchange with Sperry.

In accordance with his model of science-religion relations and of intellectual pursuits more broadly, MacKay asserts the independence of his philosophy of science and this theory of consciousness from his religious beliefs. That is, he claims that there is no direct manner in which his metaphysical beliefs legitimate his other views even though they are mutually reinforcing. Setting that claim aside for future explorations, other questions arise regarding indirect influences. For example, might religious beliefs have motivated key notions of MacKay's, such as complementarity or logical relativity?

Pursuing questions of this nature requires, minimally, comparisons of the impact of religious beliefs with the shaping influences of other factors such as scientists' research experience and their intellectual milieu. In MacKay's case, the complementarity notion originated in his encounter with a fundamental physical complementarity in the information processing systems he studied during World War II.[79] It was in this context that he first defined rigorously the notion of logical dimensionality. The logical relativity notion arose as a related development amidst discussions in postwar Britain and was stimulated in part by a related pair of papers published by Karl Popper in 1950.[80] Although such claims do not disallow the joint significance of religious beliefs, these details do portray the significance of historical patterns of development and of cumulative research experience as important dimensions of science-belief interaction.

I must leave open for future inquiry questions about the relative weight of research experience, philosophy of science, metaphysical beliefs and other factors as they shape science-belief interaction in the work of MacKay and Sperry. Their exchange on life after death does demonstrate, however, significant promise for fruitful exploration of these interactions.

In summary, the work of Sperry and MacKay as a whole offers a kaleidoscope of converging, crossing and contrasting patterns of science-belief interaction. This preliminary analysis reveals a complex dynamic of mutual influence among observation, philosophy of science, and metaphysical and religious beliefs which will richly reward further investigation.

NOTES

1. I would like to thank Vaden House, Walter Thorson, Nancy Goebel, the manuscript reviewers and the editor for their help with various aspects of this project.

2. For some helpful discussions of some issues connected with case studies in the philosophy and history of science, see, among others, W. Bechtel and R.C. Richardson, *Discovering Complexity: Decomposition and Localization Strategies in Scientific Research* (Princeton: Princeton University Press, 1993); J.H. Brooke, *Science and Religion: Some Historical Perspectives* (Cambridge: Cambridge University Press, 1991); D.L. Hull, "A Function for Actual Examples in the Philosophy of Science," in *What the Philosophy of Biology Is* edited by M. Ruse (Dordrecht: Kluwer Academic Publishers, 1989), 309-21; and E. Manier, "Social Dimensions of the Mind-Body Problem: Turbulence in

the Flow of Scientific Information," *Science and Technology Studies* 4 (1986): 16-28.

3. My objective is to provide a preliminary analysis.

4. D.M. MacKay, *Behind the Eye*, 1986 Gifford Lectures, edited by V. MacKay (Oxford: Basil Blackwell, 1991), integrates many of the key features of his work. A nearly complete bibliography is available from the author.

5. For neuroscientists a theory of consciousness addresses the relationship between neural processes and mental activities.

6. In MacKay the contrast between inside and outside stories reflects two joint features. First, it is an ontological distinction between mental events (like agency) and neural events (like neuronal inhibition). These two sets of events yield different kinds of data, different facts, which brain scientists confuse or ignore at their peril. Second, personal experience and scientific observation each provide epistemological access to the different kinds of events. The latter point is addressed by MacKay in his philosophy of science. So inside and outside data reflect both the ontological distinction (between events) and the epistemological distinction (between perspectives).

7. MacKay did not spell out his anthropology in explicit terms, but this summary pulls out relevant implications from his notions of "standpoint," "logical dimensionality," "complementarity" and "embodiment."
In MacKay's terms, the logical dimensionality of a conscious agent exceeds that of either perspective alone. MacKay's notion of logical dimensionality originates in his research on radar during the Second World War. Certain results in electronic engineering demonstrate a pattern of ontological complexity and epistemological multiplicity similar to the pattern MacKay claims is true for conscious agents. See D.M. MacKay, *Information, Mechanism and Meaning*, vol. 1, selected papers (Cambridge: MIT Press, 1969).

8. MacKay's argument supporting this claim is summarized in the discussion of logical relativity below.

9. MacKay, *Behind the Eye*, chap. 1, page 62, etc. Correlation is a relationship of correspondence for MacKay. Mental events are ontologically and epistemologically distinguishable from, but nonetheless correlated with, their respective neurological events.

10. MacKay, *Behind the Eye*, 45, italics omitted.

11. MacKay, *Behind the Eye*, 64.

12. The clearest examples of brain activity determining mental activity are the influence upon experience of drugs, electrical stimulation of the cortex and brain damage upon experience. MacKay's information flow model provides one

viable means for specifying the processes by which neural events determine mental events (sometimes called "upward causation").

13. For instance, MacKay exploited perceptual anomalies and illusions as nonredundant clues to the function of neurophysiological mechanisms. He claimed that brain science could be advanced more quickly by studying these mental phenomena than it could by studying physiological data in isolation. See, for example, D.M. MacKay, "Perception and Brain Function," in *The Neurosciences: The Second Study Program*, edited by F.O. Schmitt, G.C. Quarton, T. Melnechuk, G. Adelman and T.H. Bullock (New York: Rockefeller University Press, 1970), 303-16.

14. MacKay's work in this area has generated a diverse secondary literature, including for example, J.A. Cramer, "Science, Scientism and Christianity: The Ideas of D.M. MacKay," *Journal of the American Scientific Affiliation* 37 (1985): 142-48; C.S. Evans, *Preserving the Person: A Look at the Human Sciences* (Downers Grove: InterVarsity Press, 1977; reprint; Grand Rapids: Baker Book House, 1982); J.W. Haas, Jr., "Complementarity and Christian Thought—An Assessment: 2. Logical Complementarity," *Journal of the American Scientific Affiliation* 35 (1983): 203-9; W. Hasker, "MacKay on Being a Responsible Mechanism: Freedom in a Clockwork Universe," *Christian Scholar's Review* 8 (1978): 130-40; P.T. Landsberg and D.A. Evans, "Free Will in a Mechanistic Universe?" *British Journal for the Philosophy of Science* 2 (1970): 343-58; C.J. Orlebecke, "Donald MacKay's Philosophy of Science," *Christian Scholar's Review* 7, no. 1, (1977): 51-63; J. Szentágothai, "Downward Causation?" *Annual Review of Neuroscience* 7 (1984): 1-11.

15. The term "logical relativity" was used by MacKay in direct analogy with relativity theory in physics. In physics, temporal sequences of events differ for (are relative to) various observers who are travelling with different velocities relative to the phenomenon being observed. Similarly, MacKay argued, the knowledge of the events associated with conscious agency differs for the frames of reference provided by the inside story and the outside story (MacKay, *Behind the Eye*, 114). The original stimulus for the notion was some early work in philosophy and brain science. MacKay, *Behind the Eye*, 208.

16. Merely for the sake of argument, MacKay granted the assumption of physical determinism.

17. MacKay resisted attempts to reduce this conundrum to a simple cognitive limitation. For MacKay, the logical conclusions to this argument carry ontological, as well as epistemological, weight: limitations to the warrant of scientific predictions provide clues to the ontological complexity of the conscious agent.

18. This point reflects the correlation of mental and neural processes, an element of his theory of consciousness.

19. MacKay, *Behind the Eye*, 199. In this argument, MacKay defines the "cognitive mechanism" as the activity of the neurological subsystem that

correlates with the agent's acts of understanding and believing the prediction (192f.).

20. MacKay, *Behind the Eye*, 200. The ontological consequence drawn from the logical argument is mediated by his understanding of human nature.

21. The scientific observer will nonetheless grasp the import of the logical relativity argument and so will acknowledge that future specifications are open to the *agent*. Both the observer and the agent can be correct because they each have access to different aspects of the ontological complexity of the conscious agent posited by MacKay's anthropology. See MacKay, *Behind the Eye*, 202-4.

22. D.M. MacKay, "The Interdependence of Mind and Brain (Letter to the Editors)," *Neuroscience* 5 (1980): 1389-91; D.M. MacKay, "Our Selves and Our Brains: Duality Without Dualism," *Psychoneuroendocrinology* 7 (1982): 285-94; MacKay, *Behind the Eye*. MacKay's efforts to explicate the distinction between intralevel relations and interlevel relations are often underdeveloped.

23. MacKay, *Behind the Eye*, 63. The term "interdependence" also emphasizes the mutual determination of mental and neural processes posited by his theory of consciousness.

24. MacKay, *Behind the Eye*, 272. Embodiment is a more precise statement of the correlation of the neural and the mental. Its tentative status philosophically is a restatement of the status of "correlation" as a "working hypothesis" of neuroscientists.

25. See D.M. MacKay, "Complementarity II," *Proceedings of the Aristotelian Society Supplement* 32 (1958): 105-22; D.M. MacKay, "Complementarity in Scientific and Theological Thinking," *Zygon: Journal of Religion and Science* 9 (1974): 225-44. MacKay did not employ a distinction between ontological and epistemological features in the way done here. He preferred to talk about logical distinctions and relationships and then to assume that the reader could clearly draw out the corresponding ontological and epistemological implications of the patterns of logic being explored. In my opinion this practice leads to substantial misunderstanding of MacKay and so it is avoided here (compare, M.J. McDonald, "Exploring 'Levels of Explanation' Concepts. Part I: Interactions between Ontic and Epistemic Levels," *Perspectives on Science and Christian Faith* 41 [1989]: 194-205).

26. MacKay claimed that detached observation is essential for science because it is necessary to sustain objectivity.

27. MacKay, *Behind the Eye*, 62.

28. In MacKay's view, there are other limitations to science as well. For example, the purview of science is limited to the current space-time continuum. There is no reason internal to science why one can or cannot extrapolate the understandings of science to other domains of reality outside the current space-time domain.

29. MacKay's emphasis on logical relativity is an attempt to be precise about the nature of the conceptual frameworks which can provide adequate accounts of mind-brain relations. Logical relativity places constraints on the kinds of principles, explanations, and approaches in this arena. In this sense, the implications of logical relativity are no more debilitating for neuroscience than, say, Gödel's incompleteness theorems are for mathematics, both of which are grounded in "paradoxes" of self-reference. In fact, the results are taken by MacKay as a kind of conceptual "hygiene."

30. MacKay and Sperry used various terms depending on their own views and on the demands of their audience. These include: (a) science-Christianity (including all aspects of science and of Christianity), (b) science-theology (which focuses on the intellectual enterprise of formulating the nature of science, of religion and of their proper and improper interactions), (c) science-religion (representing an attempt to broaden the designation beyond the domain of a specific religious tradition [Christianity] and beyond the confines of academic discourse) and (d) science-belief (a subset of the previous designations which emphasizes the interactions between metaphysical commitments and science).

31. MacKay, *Behind the Eye*, x-xi.

32. MacKay's distinction between "revealed theology" and "speculative theology" in the quotation distinguishes between theological traditions which do and which do not assume the stance of the religious believer, respectively. MacKay personally advocated "revealed theology" and accepted fairly traditional notions of resurrection of the body and life after death (D.M. MacKay, "The Health of the Evangelical Body: What Can We Learn from Recent History?" *Journal of the American Scientific Affiliation* 38 [1985]: 258-65). He also understood God's creative power as undergirding the lawfulness of nature.

33. Since science depends on detached observation, it cannot completely account for the spiritual significance of events, for example. In other words, science is agnostic about theology because observation is logically relative to the experience of a believer as conscious agent.

34. See, for example, MacKay, *Behind the Eye*, chap. 12. I use "reembodiment" to suggest MacKay's notion of corporeal resurrection of a person even though the resurrected body might be significantly different from the form known prior to death.

35. The "outside perspective" of science is defined in these terms. For MacKay this ideal had normative status for all Christians since it instantiates a form of obedience to the Creator. See D.M. MacKay, "Objectivity as a Christian Value," in *Objective Knowledge: A Christian Perspective*, edited by P. Helm (Leicester: Intervarsity Press, 1987), 15-27.

36. The technical feature of information flow focused upon by MacKay in this discussion is the cooperativity of a closed loop which establishes or reflects

a new level of organization. See, for example, MacKay, "Our Selves and Our Brains"; MacKay, *Behind the Eye*, 30, 253-55.

37. For overviews of Sperry's work up through the 1980s, see R.W. Sperry, "Search for Beliefs to Live by Consistent with Science," *Zygon: Journal of Religion and Science* 26 (1991): 237-57; C.B. Trevarthen, ed., *Brain Circuits and Functions of the Mind: Essays in Honor of Roger Sperry* (New York: Cambridge University Press, 1990).

38. R.W. Sperry, "Mind, Brain, and Humanist Values," in *New Views on the Nature of Man*, edited by J.R. Platt (Chicago: University of Chicago Press, 1965), 71-92. Reprinted in *Science and Moral Priority*, by R.W. Sperry (New York: Columbia University Press, 1983), 25-44. See page 32 of the 1983 version. (References to this essay use the pagination of the 1983 version.)

39. Sperry, "Mind, Brain, and Humanist Values," 35; see also R.W. Sperry, "Commentary: Mind-Brain Interaction: Mentalism, Yes; Dualism, No," *Neuroscience* 5 (1980): 200. Revision reprinted in R.W. Sperry, *Science and Moral Priority* (New York: Columbia University Press, 1983), 77-103. Sperry does not elaborate this point in these essays and apparently contradicts it in later writings. For example, he stated recently that the complexities of spatiotemporal patterning "rule out reduction [of the laws behind macrophenomena] to lower level laws, even in principle" (R.W. Sperry, "The Impact and Promise of the Cognitive Revolution," *American Psychologist* 48 [1993]: 882; compare, 880).

40. R.W. Sperry, *Science and Moral Priority*; Sperry, "The Impact and Promise." Sperry distinguishes downward causation from "upward causation," where the latter term is taken to include causation as understood in reductive materialist understandings of science. He clearly characterizes downward causation as an interlevel relationship, but I am unable to find any characterization by Sperry of upward causation which distinguishes clearly between intra and interlevel relationships. In this sense, one cannot describe "upward causation" as equivalent to MacKay's intralevel term "causation."

41. When speaking of the relationship between mental properties and the world at large, Sperry prefers to talk about causal determination, a term which includes both upward and downward causation.

42. Sperry, "The Impact and Promise."

43. R.W. Sperry, "A Modified Concept of Consciousness," *Psychological Review* 76 (1969): 532-36.

44. Sperry, "Mind, Brain, and Humanist Values," 39-41.

45. Sperry's notions of emergence, levels of control and even "downward causation" are fairly standard. Neither Sperry nor MacKay spell out full-blown hierarchy theories, however.

46. See D.M. MacKay, "Cerebral Organization and the Conscious Control of Action," in *Brain and Conscious Experience*, edited by J.C. Eccles (New York: Springer-Verlag, 1966), 422-45, 566-74; MacKay, "The Interdependence of Mind and Brain"; MacKay, "Our Selves and Our Brains"; D.M. MacKay, Book review of *Science and Moral Priority*, by R.W. Sperry, *Neuropsychologia* 22 (1984): 385-386; MacKay, *Behind the Eye*; Sperry, "Mind-Brain Interaction"; R.W. Sperry, "Changing Priorities," *Annual Review of Neuroscience* 4 (1981): 1-15. Revision reprinted in Sperry, *Science and Moral Priority*, 104-22; also reprinted in R.W. Sperry, "Science, Values, and Survival," *Journal of Humanistic Psychology* 26 (1986): 3-23; Sperry, *Science and Moral Priority*; R.W. Sperry, "Structure and Significance of the Consciousness Revolution," *Journal of Mind and Behavior* 8 (1987): 37-65.

47. R.W. Sperry, "Mental Phenomena as Causal Determinants in Brain Function," in *Consciousness and the Brain: A Scientific and Philosophical Inquiry*, edited by G.G. Globus, G. Maxwell and I. Savodnik (New York: Plenum Press, 1976), 163-77; Sperry, "Mind-Brain Interaction"; Sperry, "Structure and Significance"; R.W. Sperry, "Psychology's Mentalist Paradigm and the Religion/Science Tension," *American Psychologist* 43 (1988): 607-13; R.W. Sperry, "In Defense of Mentalism and Emergent Interaction," *Journal of Mind and Behavior* 12 (1991): 221-46; Sperry, "The Impact and Promise." In much of his later writing a primary focus is on establishing priority for his theory of consciousness. These efforts also involve statements of principle bearing on methodology and his "working version" of a philosophy of science.

48. Sperry sees his theory of consciousness and the research on commissurotomies ("split brains") from his lab as the crucial catalyst in the turn from behaviorist theory to cognitively oriented theory in academic psychology of the United States during the 1960s and 1970s. See Sperry, "Structure and Significance"; and Sperry, "The Impact and Promise."

49. Sperry, "Psychology's Mentalist Paradigm," 607f. His usual term for his theory is a "mentalist" model of consciousness. For Sperry, "anti-realism of values" means a view of values as independent of the causal nexus of reality. His "realism of values" emphasizes the downwardly causal impact of all mental phenomena, including values.

50. Sperry, "Mind, Brain, and Humanist Values," 38.

51. Sperry, "Structure and Significance"; Sperry, "In Defense of Mentalism"; Sperry, "The Impact and Promise."

52. Sperry, "The Impact and Promise," 879. Sperry's views on free will and determinism are discussed in many publications, including "Psychology's Mentalist Paradigm"; and "The Impact and Promise" more recently.

53. From the standpoint of mentalist doctrine, as from that of common experience, one can will to do whatever one subjectively...wants *to do*. The whole process is still controlled or determined, but primarily by emergent

cognitive, subjective intentions of the conscious/unconscious mind. Thus, freedom to will our actions...is real, as are moral choice and responsibility. Yet none of these is uncaused. Uncaused behavior would be capricious, random, and out of our own control (Sperry, "Psychology's Mentalist Paradigm," 610.)

54. Sperry, "Mind, Brain, and Humanist Values"; Sperry, "Psychology's Mentalist Paradigm"; Sperry, "Search for Beliefs"; Sperry, "The Impact and Promise."

55. Sperry, "Psychology's Mentalist Paradigm," 608. Sperry names authors like Ralph Burhoe and D. Starr to illustrate liberal religion in his sense.

56. He also rejects supernaturalism because it creates a dualism which conflicts with the closed ("unified") causal system of the universe displayed by science. Neither the claims about the historical influences of supernaturalist religion nor the proposed contrast between dualism and scientific versions of causality are argued substantively, nor is historical evidence examined; these claims are simply asserted.

57. Sperry, "Psychology's Mentalist Paradigm," 48.

58. Sperry, "Psychology's Mentalist Paradigm," 46. Sperry's views clearly place a sense of God, of the sacred, and of ultimate good within an emergentist evolutionary framework. Scientific theory not only provides ultimate values, but even shapes specific theological reflection. The separation between creation and creator or the notion of life after death imply, for Sperry, a dualism which is ruled out by scientific knowledge (assumption?) of causal determinism. See, for example, Sperry, "Mind-Brain Interaction"; Sperry, *Science and Moral Priority*; Sperry, "Psychology's Mentalist Paradigm"; and Sperry, "The Impact and Promise."

59. Sperry, *Science and Moral Priority*, 49. See also "Psychology's Mentalist Paradigm."

60. Viability depends, here, upon the adoption of scientism by each religious tradition. Supernaturalism denies this preeminence to science and so stands condemned as "prescientific." This "dominance" form of compatibility between religion and science contrasts with the compatibility scheme of MacKay.

61. Sperry, "Psychology's Mentalist Paradigm"; Sperry, "The Impact and Promise." Sperry paints science as the ultimate cultural institution generated by evolution. Science provides the sole hope for overcoming our tendencies to make war, to pollute and to generate social unrest. Sperry rejects the fact-values dichotomy and subsumes all technology, including social technology, under science.

62. Strong similarities in their formulations of their theories of consciousness are acknowledged directly by both Sperry and MacKay, and

indirectly by Sperry when describing similarities between his views and the work of Dewan on control theory. At other times they emphasize their differences. MacKay, for instance, associates the notion of emergence with materialism. MacKay relies largely upon information theory while Sperry relies upon Lloyd Morgan and Gestalt psychology in his original formulation of bidirectional determination. MacKay's information theory offers clearer implications for upward causation than does Sperry's approach, and he considers Sperry's formulation of downward causation to be a bit crude. And though they share an understanding of the mental as emergent over the neural, their conceptualizations of emergence differ. MacKay, for instance, associates the notion of emergence with materialism (MacKay, *Behind the Eye*, 59). These differences may have masked the convergence of their views on the nature of mind-brain relations. See Sperry, "Mind-Brain Interaction"; Sperry, "Mental Phenomena"; MacKay, "The Interdependence of Mind and Brain"; E.M. Dewan, "Consciousness as an Emergent Causal Agent in the Context of Control System Theory," in *Consciousness and the Brain: A Scientific and Philosophical Inquiry*, edited by G.G. Globus, G. Maxwell and I. Savodnik (New York: Plenum Press, 1976), 181-98.

63. For comparison purposes I use the term "downward causation" despite MacKay's restriction of the term "causality" to intralevel processes. Both theorists agreed that empirical specification of the precise mechanisms of emergence and of mutual determination is beyond the capacity of contemporary neuroscience. The promise of their model lies in its status as extrapolation from theoretical notions which are scientifically fruitful elsewhere.

64. This difference reflects divergent assessments of human nature as an object of scientific study. MacKay's assertion of the greater dimensionality of the conscious agent in relation to the perspective of the scientific observer contrasts with Sperry's assumption that the scientific perspective can account for all ontological complexities in the agent.

65. Recall that MacKay found that mental phenomena provided data which were nonredundant with physiological data and thereby advanced his research on physiological mechanisms of sensory systems.

66. MacKay, "Our Selves and Our Brains"; and Sperry, "Mind-Brain Interaction"; and so on.

67. While Sperry posits mental activities as fully explainable within the purview of brain science, MacKay asserts that logical relativity prevents some aspects of humans from being subsumed under the perspective of a scientific observer. That is why MacKay speaks of correlations between, instead of the identity of, mental activities and higher order patterns of functional organization in brain activity. Sperry never accepted MacKay's logical relativity argument. Sperry's reaction in 1964 was that he thought that the argument might violate the assumption of a closed causal system which was necessary to science as he viewed it (see MacKay, "Cerebral Organization," 566-74). I am aware of no attempt by Sperry to counter MacKay's argument in more than a passing manner, however.

68. The phenomenology of mind-brain interaction involves, for example, experiences associated with brain damage, neurological disease and the influences of drugs as well as broader parameters of neural functioning such as sensitivity to limited wavelengths of light, to limited ranges of the temporal and spatial dimensions of events, and so forth.

69. MacKay was involved with the Neurosciences Research Program and Sperry was a key player in the "split brain" research, for example. Moreover, each of them knew of the work coming from those programs, enabling some interesting comparisons.

70. For instance, the effect of World War II on each scholar's career differed in one manner by encouraging MacKay to work on information theory. What other milieu factors might account for divergent interpretations of shared experiences?

71. Some commentators suggest that MacKay's approach might reflect some neo-Kantian elements while Sperry generally follows a more empiricist tradition. Once these family resemblances are sorted out, they might well illuminate different interpretations of common phenomenologies.

72. Sperry, "Mind-Brain Interaction"; MacKay, "The Interdependence of Mind and Brain." In his first published response to MacKay's rejoinder, Sperry makes but passing reference to MacKay's position (Sperry, "Changing Priorities," 12). In subsequent reprintings of this essay (*Science and Moral Priority*; "Science, Values, and Survival"), Sperry replaced his reference to MacKay with superficial or misleading comments. Sperry never attempted to clarify the points at issue between them or even to read much of MacKay's work. For additional details, see also MacKay, Book review; Sperry, "Structure and Significance."

73. As far as I can tell from their publications, MacKay and Sperry first met and exchanged views in September 1964 at the Vatican conference organized by Sir John Eccles (see MacKay, "Cerebral Organization," especially the discussion). Most features of their respective views had already taken shape by that time.

74. The comment by Sperry to which MacKay responded was directed at ideas of Sir John Eccles, a prominent neuroscientist and avowed dualist. Sperry, "Mind-Brain Interaction"; MacKay, "The Interdependence of Mind and Brain."

75. MacKay, "The Interdependence of Mind and Brain," 1391.

76. Sperry, "Mind-Brain Interaction," 195.

77. The direct interaction of science and belief in MacKay is confined to those instances when the object of religious belief can be independently approached by science. For example, science provides relevant, though not

exhaustive, information in the arena of human reproduction and the nature of reproduction often involves significant religious beliefs.

78. Sperry, "Mind-Brain Interaction."

79. MacKay, *Information, Mechanism and Meaning*, chap. 1. MacKay's application of the complementarity notion to science-Christianity relations also dates from this era. Sorting out the question of influences here would also require identification of the roots of his anthropology.

80. See MacKay, *Behind the Eye*, 208. As for Sperry, prior to the results of "split-brain" studies on humans, he was likely to avoid any kind of "mentalist" notions, including values. So he attributed significant weight to that particular research experience (Sperry, "Search for Beliefs").

13

Control Hierarchies—A View of Life

David L. Wilcox

INTRODUCTION

Hierarchy theory is a concept referring to a class of theories that conceive of reality in terms of levels of organization. It is best known as the view that sees reality in terms of wholes that are physically composed of parts. For instance, skipping a few levels, the idea is that the universe is composed of galaxies, a galaxy is composed of planetary systems, a planet is composed of ecosystems, an ecosystem is composed of organisms, an organism is composed of cells and a cell is composed of molecules.

There seems to be a paradox in our knowledge of biotic "self-organization": the higher or more inclusive levels of physical structure control the activities of the lower levels, yet the inclusive levels are constructed by the lower levels. In this paper, I show that this is a paradox only for certain types of hierarchy theory, such as hierarchies of physical components, but not for control hierarchy theory. However, control hierarchy has not been generally accepted. I show that this is due to a conflict between the central ideas of encoded information and specified goals. These are also needed for an empirically adequate control hierarchy and methodological as well as metaphysical naturalism.

The development of control relations in an organism exemplifies the paradox. In the adult organism, the state of lower levels is measured, evaluated and modified by control systems dedicated to maintaining the normal system values set by the higher levels. For instance, hypothalamic nuclei (brain centers) measure the osmotic pressures of body tissues, and send commands via neuronal pathways to turn on or off the gene for the hormone that in turn controls the kidney's reabsorption of water and, thus, osmotic levels.[1] Thus, organ systems constrain the action of their tissues, and control is top down, for example, holistic.

But how does this top-down control develop? Complex organisms start as undifferentiated single cells and develop *through* the levels of complexity. At each level, new structural complexity appears through characteristic interactions of existing lower-level components. Although simple tissues interact to produce complex organs, paradoxically the organ being formed somehow seems to constrain the interacting tissues. Paradoxically, the whole seems both determined and not determined by the parts.

As an example of such "controlled emergence," consider the optic cup. This is an embryonic extension of the brain which forms the retina. The optic cup releases a chemical message (termed an inducer) which directs the overlaying epithelium to thicken into a lens. If the optic cup of a large-eyed species of salamander (*Ambystoma tigrinum*) is used to induce lens formation in the epithelium from a small-eyed species (*Ambystoma punctatum*), the two tissues interact to form an eye of intermediate size and normal morphology.[2] The proliferation values of the mismatched tissues are modified to match an organ-level specification for eye morphology. *Organ*-level specifications thus seem to direct tissues to precise goals—despite the fact that the organ itself is still under construction.

Obviously hierarchical structure is involved in the control and emergence of biological complexity in organisms. However, which model of hierarchy is adequate for understanding that emergence? In this paper I will evaluate the empirical adequacy of four models of biological hierarchy. These are the hierarchies of classification, composition, information and control. My conclusion is that only hierarchies of control incorporating goal-seeking behavior are adequate to describe organisms. However, since material goal-seeking systems such as organisms cannot generate their own goals, the origin of the goals must be sought in nonmaterial reality. I conclude that this metaphysical implication is the reason why goal-seeking control hierarchies are rarely employed: it conflicts both with methodological and with metaphysical materialism.

HIERARCHIES OF CLASSIFICATION

Discussions of biotic hierarchy first arose in the eighteenth century in the midst of debates concerning classification and comparative anatomy. These initial formulations (such as the higher anatomy of German *Naturphilosophie*) included a sort of "semi-Platonic" hierarchy of immaterial "archetypes" which shaped material biotic reality, controlling both embryonic and phyletic development.[3] This view involved the metaphysical belief that material biotic entities reflected a fundamental immaterial reality. This belief was rejected by positivism, which came to dominate science.[4]

Indeed, the choice of a specific commonality as "fundamental" makes hierarchies of classification acute indicators of worldviews. The nineteenth-century shift in the basis for taxonomic classification demonstrates this. In a worldview which assumed design, the *fundamental* reality was that design. Hence classification was by archetype—even in the work of men like Richard Owens who accepted common descent. If, however, the material world is believed to run without a designer's hand on the helm, the common ancestor and the autonomous forces of change determine the present form of the organism. Thus, phylogeny is not just history, it is creative cause. Under a materialistic consensus, the common archetype became the common ancestor, and the forces producing variation in the context of common ancestry shifted to autonomous material happenstance rather than the free will of God.

Such alternate taxonomies are logically equivalent, since they are based on a choice of which common descriptive element is most important (for example, common ancestry or common archetype). Thus, hierarchies of classification can only describe biotic emergence. They cannot explain biotic emergence, for they use it as a criterion for classification.

PHYSICAL COMPONENT HIERARCHIES

A model widely used to explain the emergence of biological structure is the material composition hierarchy—ecosystems made of organisms, organisms composed of organs, organs of tissues, tissues of cells, cells of organelles, organelles of molecules, etc. Such a view has been assumed as the "natural" view under the materialist hegemony of Western thought. Organisms are assumed to have emerged by combining simpler parts into more complex entities (cells into metazoans). If organismic phenomena are reducible to molecular events, the atomic level holds the keys to understanding reality.

Certainly, living things are physically constructed of smaller units. But such a model is inadequate to explain the role of information in determining the complexity of the whole entity. Either such complexity must simply *emerge* from the properties of its parts, or it must be applied from the context of the parts. Composition hierarchies explain neither case.

If physical or composition hierarchies are to account for the emergence of complexity, they must be able to account for the origin and role of the information required for the emergence of complexity. One possible origin of the information is in the parts of a physical component hierarchy. D.R. Brooks, J. Collier, B.A. Maurer, J.D.H. Smith and E.O. Wiley term the complexity of material entities "bound information," by which they mean "a measure of the causal power of a physical system in the sense of its maximal ability to make a distinction between, or a difference in, the systems with which it interacts."[5] Such bound information is more or less equivalent to the measurable structural complexity. Since they are also material entities, components have measurable bound information. The question concerning component hierarchy is: Is the information bound to the components a property of the components or of the entire physical component hierarchy?

In some hierarchies of physical composition, the physical properties of the components dictate the structure of the complete entity. They can self-assemble. For instance, the proteins of a mitochondrial respiratory assembly or a chloroplast quantasome will fit together in lock-and-key fashion and self-assemble into the larger structure. However, such composition hierarchies cannot explain emergence. The idea of the self-assembling system starts as a flawed analogy with nonbiological systems such as crystal growth. The spontaneous formation of a respiratory assembly depends on at least eight proteins having *exactly* the correct shapes and charges. Those proteins are effective material causes because they have very *high* levels of bound information. However, the twenty amino acids from which they are made have very *low* levels of bound information, and the chemical compounds from which the amino acids are made have almost *no* bound information. At the level of the small molecule (H_2O, CO_2, etc.), one finds components which are not *specified products* assembled from entities under the direction of some biotic blueprint. The material constraints exercised *by* the respiratory proteins reflect information added during protein construction, transferred from the genetic blueprint (codon sequence). Therefore the information cannot originate from the parts of a component hierarchy. In applying material composition hierarchies to living things, the constraints are ultimately not physical complexity but specified complexity (information).

So the information cannot originate in the parts because the simpler the parts, the smaller the amount of bound information. Neither can the information originate in the completed whole, because it does not yet exist. Could it originate in the emerging whole and be provided by so-called morphogens? For instance, in insects, a simple gradient of signal chemicals causes the "emergence" of highly differentiated segments. Thus a very simple context seems to "cause" a complex product.[6] However, in a different insect (or a vertebrate) the same chemical gradient produces a completely different body. Therefore, the constraints which shape bodies are present in the responding cells, not in the signal gradient. The meaning of the signal gradients is *defined* within the responding cells, and thus the body pattern which constrains the cells must be *within* them, perhaps encoded as a sort of genomic "blueprint." Biotic components are indeed linked to form larger entities, but the information that controls the linking entities is neither in the components nor in the whole. It originates from the genetic blueprint present in every cell.[7]

INFORMATION-BASED HIERARCHIES

Physical component hierarchies are empirically inadequate to account for the emergence of complexity because physical components are the medium, but not the source of information. Therefore a promising alternative model would be a hierarchy of *information.* Such a hierarchy in organisms is analogous with human language. Genes are "words" (messages) encoded on DNA (the medium) which describe aspects of the organism. The first level "alphabet" is the codon level, sequences of three nucleic acids (for example, GAG) with assigned meanings—each triplet signifies a particular amino acid. Thus, a sequence of codon messages specifies a *sequence* of amino acids.

Proteins are strands of amino acids (twenty kinds), arranged in a specific sequence. Thus, the *encoded* information of the codon message is expressed as the bound information of the protein. Such messages encoded by codon sequence are termed genes. Protein strands automatically coil and fold into three-dimensional objects with highly specific shape and chemical properties, which may act as building blocks—or as complex machines. Since the amino acid sequence determines that folding, the codon message specifies the nature of the finished protein.

So, we already have two levels of "language." Codons are assigned meanings such as "leucine" (an amino acid), whereas genes (codon sequences) are assigned meanings such as "hemoglobin" (a protein). The second level alphabet—the set of structural genes—is assigned information which implies the chemical structure, activity and control properties of the proteins. This gene "alphabet" is then used to specify and control simple organelles (cell "organs") such as ribosomes, nucleosomes, or respiratory assemblies.

The third level of specified meaning on the DNA uses the structural genes as its alphabet to encode messages. It does so through the use of "control" genes, specific recognition sequences of DNA attached to structural genes, and unique proteins which can identify specific sequences.[8] A set of structural genes controlled by such a regulatory protein therefore constitutes a third-level message. The "meaning" of the gene which makes the regulatory protein is the specific array of structural genes tagged with sequences which the protein can recognize and thus activate.[9] Again, the definition of the regulatory gene has *nothing* to do with its intrinsic bioactivity. It can not *directly* recognize any gene.

Finally, since regulatory (or recognition) genes are defined by their ability to control groups of structural genes, the way is open to have regulatory genes which control other regulatory genes, thus producing fourth, fifth or sixth level messages.

Thus, we have established that the information hierarchy found encoded on the DNA of organisms forms a nested instruction set which describes the structure of the organism in a hierarchical fashion. The codon stands for an amino acid, the structural gene describes a protein's initial structure, and a regulatory gene specifies the genic activities needed to produce aggregates and organelles. Above these levels, although the hierarchy of biological description becomes much harder to follow,[10] the analogy can be extended to suggest that the "N[th]" level of regulatory gene specifies (is a blueprint for) even complex behavior such as star charts in migrating birds or bridge building in foraging army ants. This may seem unwarranted speculation; however, the material complexity of organelles, cells, tissues, organs, organisms and hives is so great that it must be the realized product of detailed descriptive blueprints encoded somewhere, because this complexity is inherited and its pattern of inheritance is predictable.

Information-based hierarchies account better for empirical reality than physical component hierarchies. The main reason is that physical components can function only as carriers of information, but not as information itself. Furthermore, information-based hierarchies can account for descriptions of organic structures and physiological states. Such descriptions are given in terms of a descriptive hierarchy of specifying messages encoded on the DNA. Moreover, information-based hierarchies can account for prescriptions of the normal states of control mechanisms in terms of a description of normal system states. For instance, the protein hemoglobin is designed to release oxygen maximally at that specific oxygen tension characteristic of a stressed tissue, thus maintaining the normal high tissue level of oxygen. The specific affinity curve for hemoglobin—which depends on the codon message—thus defines the oxygen level which "should" be maintained in the tissue. Finally, information-based hierarchies can account for the connections required for control relations in terms of information flow.

The importance of information in organisms is that no information chain can exist without a mental origin. The fact that no information can exist without a mental origin, a will and a purpose has put heavy pressure on the materialistic consensus.[11] The orthodox response has been to assume that such packages of information as organisms are simply maps of *material reality* which somehow branched off from sufficiently complex material systems at an early date. If that was so, if the information is just "hardware" specifications, then organisms' unique qualities can still be viewed as emerging spontaneously from the material "ground of all being." However, the material biological world reflects the function and form of its encoded blueprints. Hierarchies of information are a necessary condition for the existence of organisms.

However, information-based hierarchies cannot account for organic processes of control. Instead, a model of a hierarchy of controls is needed, for living things are dynamic, changing entities, engaged in active and interactive governance, maintaining homeostatic levels and producing new complexity. But, a model of a hierarchy of controls requires more than a description of the organism, or even a prescriptive statement of the normal state of the various material components of the organism. The central concept of control is that there is an actor such as an organism which realizes certain goals. The goals and the control mechanisms may indeed be specified by messages on the DNA strand, but the messages are not enough. The specifying message must be interacted with, even obeyed, by component actors, (even if the components have been previously constructed—as specified by that message.)

HIERARCHIES OF CONTROL

Therefore in this section, I will describe the key characteristics of control hierarchies in order to show that they are empirically adequate to account for living organisms as goal-seeking systems. To account for goal-seeking behavior, a control hierarchy requires controlling entities and a way of building them into a control hierarchy, a hierarchical arrangement of tasks, different kinds of hierarchies integrated into a control hierarchy, a mechanism for the flow of information and for the development of new control relationships.

What kind of entities make a control hierarchy? Hierarchies of physical composition are made of material objects. Linguistic hierarchies are made of words. Developmental hierarchies consist of events. The common characteristic of these hierarchies is that they are made of concrete entities. A hierarchy of control, however, consists of abstract goals, tasks and operators.[12] The concrete entities of the hierarchies of physical composition and of information-based hierarchies serve to assign abstract tasks to other concrete components operating under a central goal which they are called to realize.[13] Therefore, the unit of a control hierarchy has both abstract and concrete aspects.

The goal-directedness of the entities making up a control hierarchy is not a *boundary* constraint, a fence to corral the molecular herd; it is a *central* constraint—a core definition of what that entity is to be like. This means that the goal of entities is not what they can be forced to do by exterior pressure, but rather what they are intended to do if they follow their specified design. Since component designs are parts of the larger design of the whole organism, their goals are integrated under the goals of the whole organism.

A further characteristic of the basic or determining entities of a control hierarchy is that they have the most complexity, whereas the entities of physical component hierarchies have the least. This follows from the fact that the basic entity of a control hierarchy possesses (1) definitions of the goals (prescriptive information), (2) entity status evaluation (descriptive information), (3) error measures (evaluative information), (4) corrective processes (reaction norms) and (5) decision-making capacity. These characteristics are required for the entity to be able to empower other components to the realization of particular goals.

Goal realization requires tasks to be assigned to sub-entities. Thus the levels of a control hierarchy are entities with assigned tasks. Larger entities are indeed composed of smaller components, but those components are specified and formed to perform a task for the larger entity. Larger tasks can indeed be broken into smaller tasks, but those larger tasks shape the components. In a control hierarchy, each level's task is part of a larger task—and the parts are "designed" (assigned) for the purposes of the whole.

The *dynamic entities* making up a control hierarchy make possible the construction of complex networks in which interactive flow of information can take place. The describing informational units (genetic specifications) and the specified components described (such as proteins) are caught up in a dynamic, goal-seeking loop. Since proteins can be specified which will be able to locate and activate other genes, and, at the same time, act as chemical sensors of their "environment," it becomes possible for the descriptive DNA information hierarchy to adaptively govern growth and maintenance via a network of control relations. This contributes to the explanatory power of control hierarchies.

For instance, control hierarchies have the advantage of accounting for complexity in terms of dynamic process. Thus one can avoid referring to a mysterious vital substance or homunculus as the source of complexity. Rather, the complexity is in a network of information flow. Descriptive or prescriptive information may be encoded as linear sequences, but tasks must be encoded in networked relationships. For instance, homeobox genes are sequential sets of regulatory genes which read the chemical gradients released in embryos, and control linear differentiation and segmentation. Groups of homologous (even cross-reactive) homeobox gene sequences have been shown to code for DNA sequence recognition proteins (helix-turn-helix), and have been found in *Drosophila*, man, mouse,[14] nematode,[15] maize[16] and even yeast.[17] However, sequence homology and cross reactivity show that homeobox genes do not *contain* the information they control. If the same gene product controls antenna production and vertebrate brain differentiation, homeoboxes are simply obedient parts of certain levels of various control hierarchies which are incorporated into the genetic network. As Tim Beardsley put it: "Smart genes, it seems, talk to one another a good deal during development. The process continues, fractal-like, at successively smaller scales, eventually elaborating stripes of protein expression."[18] A. Garen concludes a recent paper with the following sentence: "The elusive homunculus is finally shaping up not as an object but rather as a dynamic process."[19] However, Garen's conclusion is incomplete. The homunculus is more than an object or a process. It is a entity with activated information encoded in the network of relationships between the *Drosophila* control genes. Development only reveals—or realizes it.

Is the coding capacity of genetic networks sufficient to encode the tasks of a biotic control hierarchy? Stuart A. Kauffman has estimated the capacity of information flow in biological systems.[20] He estimates that with 100,000 genes, there are at least $10^{30,000}$ possible patterns of expression. Any pair of these loci could be part of a feedback loop (either negative or positive), and the number of such potential loops increases exponentially.[21] In their discussion of genetic control,

S. Brenner, W. Dove, Ira Herskowitz and Richard Thomas state that the number of stable steady states for a network is 2^m, where m is the number of direct or indirect positive feedback loops.[22] Clearly, the genome has the same potential for interactive information flow as the brain itself. So then, the genome has the capacity to encode the tasks of the biotic control hierarchy.

Control hierarchies have the potential to account for the evolutionary emergence of new control relations in terms of the modification of existing ones and the construction of new ones. Effective mutations would change the networking potential of the mutated genes. For instance, an existing control relation could be inhibited or facilitated when a recognition protein is modified due to mutation. This suggests that over multiple generations, new (transformed) control relations may emerge between existing nodes in the network by mutation and selection of the nodes. New (*de novo*) control relationships could also arise via the attachment of new control sites to additional gene loci. This would be another "mutational" process entirely, requiring the duplication of gene sequences and the rearrangement of chromosomes—without disrupting the functioning genetic network. Currently, there is no known mechanism by which new control loops could arise gradually between mutated genes. Thus the origin of new control relations via this particular mechanism remains a mystery. This, however, is due not to a limitation in the explanatory power of control hierarchies, but to the lack of one of two possible mechanisms for the production of new control relations in a control hierarchy.

On the other hand, the appearance of new control relationships does not imply that random mutations can generate new *goals* for components in a control hierarchy. A new goal would require a new control loop. B. Borstnik, D. Pumpernik and G.L. Hofacker recently suggested that point mutation can act as an optimal search process,[23] and A.A. Pakula and R.T. Sauer state:

> It is an especially encouraging result that proteins appear to tolerate most substitutions, even those that are destabilizing, without significant changes in the native structure. For proteins whose structures are known, this means that it is reasonable to interpret mutant phenotypes in terms of the wild-type structure.[24]

Thus, mutations only have meaning in the context of their original goals. They are changes in existing relationships, not newly created relationships.

The empirical adequacy of control hierarchies also manifests itself in accounts of the control of embryonic development. An example of how control hierarchies act in development is what G.P. Wagner refers to as the "individuation" of specific units of morphology.[25] Individuated structures look like individual entities with cells "committed" to organ formation, for they respond as single units to changing environmental signals or genetic changes (mutations). Wagner refers to the "emergence of autonomy" in a group of cells committed to develop a particular organ, commenting that, "Anatomy emerges at the level of the organ but not at the level of the parts."[26] Thus, an individuated structure is composed of components which are acting under the direction of a set of coordinated higher-level control norms. It is this set of goals which determine the outcome of the individuated structure, not its environment. This underlying hierarchy of control is the central essence of development, and only the control hierarchy is an adequate model. Individuation is a goal-seeking mechanism.

Thus, from the point of view of a control hierarchy, there is no paradox in the development of higher from lower levels even though the former controls the latter, for the full controlling hierarchy of goals and tasks is present even in the zygote. As the embryo develops, interacting tissues simply realize the organ-level goals already present. In conclusion, since organisms are goal-seeking systems,[27] and since control hierarchies specify *goals* and enable components to fulfil those goals, control hierarchies are empirically adequate to describe organisms.

THEORY CHOICE AND WORLDVIEW COMMITMENT

If goal-seeking control hierarchies are such a powerful key to understanding organisms, why have they been so rarely utilized? I suggest this is because this sort of control hierarchy theory does not ignore the fact that organisms are information-carrying and goal-directed systems. Information-based control hierarchy theory refuses to explain these facts in terms of material causes. Ultimately, this raises the spectre of the mind as an encoder and director. Any theory of organism, whether hierarchical or not, that explains organisms in terms of an encoder and director is unacceptable in a climate of materialism. With materialism, the emergence of organic complexity and goal-direction must be explained in terms of material causes acting without direction.

Olivier Rieppel has suggested that,

The attempt of evolutionary theory to capture the biological world in a single and unified causal explanation rests on a reduction of the four Aristotelian causes to one (efficient)—a heritage of the Enlightenment, which...hampers a full understanding of the complexity of living systems."[28]

As he put it,

According to the atomistic perspective, organisms were formed by an aggregation of parts or particles....As these parts were in principle interchangeable, organisms were conceived as fundamentally variable...what is therefore required is a "generative structuralism" governing not only the invariance of structures, but at the same time their generation and transformation....[29]

Thus, Rieppel feels that morphogenetic "generative principles" dictate the possibilities of biological form.[30]

However, Rieppel is not proposing a theistic explanation, which again shows that the materialist rejection of goal-oriented control hierarchy is unwarranted. Hence, the neglect of such structuralist insights suggests a high degree of "teleophobia" in biology. Investigators seem to feel they must defend themselves against an accusation of teleological thinking as if it were a charge of moral turpitude.[31] Von Brucke (as quoted by systems theorist M. Mesarovic) described teleology "as a lady without whom no biologist can live—but he is ashamed to be seen with her in public." Mesarovic continues: "For an important class of situations, one can develop an effective specification of the system only if one is using a goal-seeking (that is, teleological) description."[32] Such a reductionistic fervor has long separated the insights of physiological functional analysis from phylogenetic morphological analysis.[33] The neglected issue is not the search for biological order within the constraints imposed by the goals of physiological systems, it is the search *for* these constraining goals. Only a fervent desire would keep looking for them within the material hierarchy.

The difficulty for the materialist is that a goal-directed material system is unable to generate its goals internally.[34] A model which presumes goals and tasks sounds suspiciously like it has been derived from theology. It may indeed be true that goal-directed systems are congruent with a faith in a rational designer. However, control hierarchies should be evaluated on their ability to match the structure of biological reality, irrespective of such religious congruence—or its lack. I thus argue that control hierarchies are empirically more adequate to account for the functioning of organisms than any of the other hierarchies discussed.

The superiority of control hierarchies is being reluctantly accepted. In reviewing *The Logic of Life*, a recent book supporting a new holistic control perspective, K. Schmidt-Nielsen said, "Physiology will be indispensable in putting together and interpreting the masses of detailed information emanating from the revolutionary progress in molecular biology."[35] In a theistic climate, that insight would have been obvious long ago.

CONCLUSIONS

I have reviewed the empirical strengths and metaphysical implications of four models of hierarchy. Hierarchies of classification, although they reveal the metaphysical commitments of their users, have no power to explain emergence. They simply are a process of grouping things by fundamental likenesses, in which "fundamental" is due to metaphysical assumptions. Hierarchies of material composition fail to explain the emergence of complexity and information. The paradox of higher-level material controlling entities which seem to be constructed by the very entities which they control is explained by information-based control hierarchies. Thus, the tenacity with which material hierarchies have been held in the face of empirical inadequacy must be due to their metaphysical "orthodoxy" within materialism. Still, hierarchies of information, although they have the characteristics needed to store the specifying information for making components at all levels, cannot encode or decode themselves and cannot obey their own instructions. Only a control hierarchy composed of entities which read and follow the genetic instructions, fulfilling the tasks of the organism as a whole, has the empirical power to model biological reality. And that model clearly is at home in Yahweh's universe.

Neo-Darwinian processes can not form genetic control networks—either by "orthodox" step by step change or by "heterodox" macromutational leaps. Modern genetics must explain new networking through randomized cross-over to reattach reading codes. Perhaps it could happen that way—in Yahweh's world, where he governs the spin of every electron. But in Monod's universe of chance and necessity, such an explanation for magnolias, pangolians, butterflies and scientists requires a Kierkegaardian leap of faith of its "true believers."

NOTES

1. ADH—Antidiuretic hormone.

2. P.J. Bryant, and P. Simpson, "Intrinsic and Extrinsic Control of Growth in Developing Organs," *Quarterly Review of Biology* 59 (1984): 387-415.

3. A. Desmond, *The Politics of Evolution* (Chicago: University of Chicago Press, 1989).

4. J.R. Moore, "1859 and All That: Remaking the Story of Evolution-and-Religion," *Charles Darwin, 1809-1882: Centennial Commemorative* (Wellington: Nova Pacifica, 1982), 167-94.

5. D.R. Brooks, J. Collier, B.A. Maurer, J.D.H. Smith and E.O. Wiley, "Entropy and Information in Evolving Biological Systems," *Biology and Philosophy* 4 (1989): 407-32.

6. A. Garen, "Looking for the Homunculus in Drosophila," *Genetics* 131 (1992): 5-7.

7. E. Vrba and N. Eldredge, "Individuals, Hierarchies and Processes: Toward a More Complete Evolutionary Theory," *Paleobiology* 10, no. 2 (1984): 146-71; and E. Vrba and S.J. Gould, "The Hierarchical Expansion of Sorting and Selection Cannot be Equated," *Paleobiology* 12, no. 2 (1986): 217-28, propose that aggregates of parts can become integrated systems through the interaction of two material hierarchies. This solution for the problem of emergence fails for the same reason as other material hierarchies fail: it neglects the fact that the components in both hierarchies are specified by encoded information.

8. B.W. Matthews, "No Code for Recognition," *Nature* 335 (1988): 244-45.

9. An example of such third-level messages are the homeobox genes which control segmentation in many creatures. If they are mutated, fruit fly sections of the body act as the "wrong segment"—for instance, fruit fly head sections may make legs for antenna.

10. R.F. Weaver and P.W. Hedrick, *Genetics* (Dubuque: Wm. C. Brown, Publishers, 1989).

11. W. Gitt, "Information: The Third Fundamental Quality," *Siemens Review* 56, no. 6 (1989): 2-7.

12. Talbot, a physiologist, suggested that systems theory adds two specific useful principles (a) the concept of "task", where "task or purpose here is not intent, but design objective deduced from performance," and (b) the concept of system components as "operators" assigned to such tasks, where "a system is conceived as comprised of a network of such elementary operations performed by its subsystems." Talbot phrased this in a reductionistic manner, but clearly it is more natural to view the "tasks" in relationship to assignments by the whole system as a single entity. (A. Talbot and U. Gessner, *Systems Physiology* [New York: John Wiley and Sons, 1973]).

13. Hierarchies of control incorporate a variety of other hierarchies including hierarchies of system state descriptions, of system "health" status, and of reaction mechanisms.

14. J. Marx, "Homeobox Genes Go Evolutionary," *Science* 255 (1992): 399-401.

15. C. Kenyon and B. Wang, "A Cluster of Antennapedia—Class Homeobox Genes in a Nonsegmented Animal," *Science* 253, no. 2 (1991): 516-17.

16. J. Rennie, "Homeobox Harvest," *Scientific American* 264 (1991): 24.

17. G. Riddihough, "Homing in on the Homeobox," *Nature* 357 (1992): 643-44.

18. T. Beardsley, "Smart Genes," *Scientific American* 265 (1991): 86-95.

19. A. Garen, 5-7.

20. S.A. Kauffman, "Antichaos and Adaptation," *Scientific American* 265 (1991): 78-84.

21. As Kauffman points out—3^n where n is the number of loops.

22. S. Brenner, W. Dove, I. Herskowitz and R. Thomas, "Genes and Development: Molecular and Logical Themes," *Genetics* 126 (1990): 479-86.

23. B. Borstnik, D. Pumpernik and G.L. Hofacker, "Point Mutations as an Optimal Search Process in Biological Evolution," *Journal of Theoretical Biology* 125 (1987): 249-68.

24. A.A. Pakula and R.T. Sauer, "Genetic Analysis of Protein Stability and Function," *Annual Review of Genetics* 23 (1989): 289-310.

25. G.P. Wagner, "The Biological Homology Concept," *Annual Review of Ecology and Systematics* 20 (1989): 51-69.

26. Wagner, 51-69.

27. M.D. Mesarovic, "Systems Theory and Biology—View of a Theoretician," in *Systems Theory and Biology: Proceedings, Systems Symposium 3* (New York: Case Western Reserve University Systems Research Centre/Springer-Verlag, 1968).

28. O. Rieppel, "Structuralism, Functionalism, and the Four Aristotelian Causes," *Journal of the History of Biology* 23, no. 2 (1990): 291-320.

29. Rieppel, 291-320.

30. Rieppel, 291-320.

31. For example in W.A. Mitchell and T.J. Valone, "The Optimization Research Program: Studying Adaptations by Their Function," *Quarterly Review of Biology* 65, no.1 (1990): 43-52.

32. Mesarovic.

33. G.V. Lauder, "Functional Morphology and Systematics: Studying Functional Patterns in a Historical Context," *Annual Review of Ecology and Systematics* 21 (1990): 317-40.

34. Gitt, 2-7.

35. K. Schmidt-Nielsen, "Unifying Science," *Nature* 367 (1994): 230.

14

The Concept of Hierarchy in Contemporary Systems Thinking— A Key to Overcoming Reductionism?[1]

Sytse Strijbos

THE CONTROVERSY BETWEEN REDUCTIONISM AND ANTIREDUCTIONISM

Broadly speaking, the controversy between reductionism and antireductionism concerns the question of reducibility in the sciences.[2] This question of reducibility has a long history in biology, in particular where the issue is whether the phenomena of life can be accounted for adequately in terms of physics and chemistry. The question of reducibility has always been important in psychology, too, where it bears directly on the well-known problem of the relation between the psychical and the physical, between mind and body. Here a reductionist perspective implies that it is possible, in general, (ultimately) to reduce all mental phenomena to their biological and physical-chemical substrata.

Where do the systems sciences stand in this controversy between reductionism and antireductionism? Often indeed the scientific developments of recent decades in this field—cybernetics, general systems theory, autopoiesis, the paradigm of self-organization, etc.—have been greeted as victories over a sterile reductionism that avoid the trap of a vague and scientifically unfruitful antireductionism. Playing a fundamental role in this connection is the concept of hierarchy featured in the works on systems theory written by von Bertalanffy, Laszlo, Jantsch and others. Jantsch, in his book *The Self-Organizing Universe*,[3] speaks of a "multilevel dynamic reality" in which "each new level brings new evolutionary processes into play which co-ordinate and accentuate the processes at lower hierarchical levels in particular ways." And therefore, so this writer continues, "reduction to one level of description is never possible."[4] Systems thinking overcomes the "ruling reductionism of Western science whose aim is to reduce all phenomena to *one* level of explanation—a level which physics hopes to find in the microscopic, in the basic structure of matter."[5]

Yet in the controversy between reductionism and antireductionism there are also systems theorists who explicitly adopt a position at the pole of reductionism. One can refer, in this regard, to Herbert Simon's work in systems theory and cognitive psychology. While von Bertalanffy and others introduce the concept of hierarchy in the conviction that they will thereby be able to offer resistance to a reductionist approach in science, Simon develops a general theory of hierarchical systems precisely because he is a reductionist in principle.

How is one to account for this striking contrast between von Bertalanffy and Simon? Is it indicative of a certain ambiguity in systems thinking with respect to reductionism? Or must one perhaps conclude upon closer examination that von Bertalanffy's resistance to reductionism offers no obstacle to a broader form of reductionism? This paper analyzes von Bertalanffy's and Simon's concepts of hierarchy.[6] We consider the positions they adopt in the controversy between reductionism and antireductionism, seeking thereby to find an answer to the questions that have been posed and, moreover, to bring to light any "metaphysical beliefs" that may be connected with their concepts of hierarchy.[7]

THE CONCEPT OF HIERARCHY IN VON BERTALANFFY

Next to the concept of the "open system," the concept of "hierarchy" occupies a central place in von Bertalanffy's thought.[8] As a biologist he is deeply impressed by the magnificent architecture of the organism and of all living nature. Thus he devoted a separate chapter to this subject in his *Problems of Life*, a chapter that begins:

> We find in nature a tremendous architecture, in which subordinate systems are united at successive levels into ever higher and larger

systems. Chemical and colloidal structures are integrated into cell structures and cells, cells of the same kind to tissues, different tissues to organs and systems of organs, these to multicellular organisms, and the last finally to supra-individual units of life.[9]

The splendid architecture that we observe in the organism is typical of a pattern that we encounter in many fields besides biology, so von Bertalanffy believes, including, for example, psychology and sociology. Hierarchical organization is the ordering principle even for total reality. And therefore: "A general theory of hierarchic order obviously will be a mainstay of general systems theory."[10]

Von Bertalanffy mentions in the same breath the architecture that we observe in the organism and the hierarchic order of the cosmos. However, a statement about the structure of the organism or about the universe do not have the same status. The reason is that we cannot experience the cosmos in its totality because we are part of it. We have no standpoint outside it. To move from the hierarchical nature of a part of reality, a system, to a hierarchical structure of total reality, is an extrapolation. When von Bertalanffy and other systems thinkers assert that in the light of systems thinking reality discloses itself to us as a "tremendous hierarchical order of organized entities,"[11] then one must remember that this statement involves a particular way of looking at reality. "We presently *'see'* the universe as a tremendous hierarchy," von Bertalanffy correctly remarks.[12] For, as said, it is beyond our capacity to experience the universe, that is, the cosmos in its totality, and the view that it is a hierarchy of systems accordingly exceeds the insights and results of all sciences. This conception has the status of philosophical ontology or cosmology, and von Bertalanffy calls it systems ontology. "This is *systems ontology*—what is meant by 'system' and how systems are realized at the various levels of the world of observation."[13]

Because ontology involves a particular way of "seeing" reality, the question arises concerning how such a view of total reality is to be justified. On what is it based? That ontological or cosmological positions exceed the insights of science does not imply that they have no bearing at all upon scientific questions and discussions in various fields of research. On the contrary, via the fundamental discussions in one's field, the researcher is stimulated to place his or her discipline in a broader and indeed ontological context. So too von Bertalanffy. As a theoretical biologist he was confronted by two main schools of thought in his field, mechanism and vitalism. Mechanists defended a theory based on the machine and sought to reduce all the phenomena of life to physical or chemical regularities, while vitalists maintained that the living organism is more than a machinelike entity and alluded to a mysterious and unexplainable life force, an irreducible factor X. But what that "more," that factor X is, remains vague in the defenders of the vitalist legacy.[14]

The questions that confronted von Bertalanffy in the debate about the foundations of biology compelled him to take a position with respect to the controversy between reductionism and antireductionism. In my judgment, the choice that must be made here is between two fundamental postulates: either one proceeds from the idea of the mutual irreducibility, which can be called the postulate of discontinuity, or one adopts the contrary position and postulates continuity. These postulates are of much broader importance than for biology alone. Everyone who theorizes or philosophizes must accept one of them as a basic ontological position. The matter is one of an *a priori* choice, or metaphysical belief, if one prefers to call it that.

Behind the *a priori* choice there is still something else that transcends human experience, namely, a conviction regarding the possibility of grasping the whole of reality in one theoretical view. To proceed from the idea of the mutual irreducibility of the different fields of reality is to say implicitly that such a theoretical total view on the whole of reality is *not* within human reach. To postulate continuity means the opposite: a theoretical grasp of the whole of reality *is* possible by reducing the diversity in reality to a common denominator.

Thus it is of interest to see how von Bertalanffy positions himself with respect to reductionism and antireductionism. He seems to speak out very clearly against physicalistic reductionism "that considers physical phenomena as the sole standard of reality."[15] He seems no less clearly to oppose other kinds of reductionism as well. That is, he seems to reject the idea that any one scientific discipline, whichever it might be, would be the most fundamental in truth. In defending the independence of biology he therefore at the same time guards against falling into the trap of biologism, "that is, into considering mental, sociological and cultural phenomena from a merely biological standpoint."[16] I have deliberately used the word "seems" several times. For how can one rhyme von Bertalanffy's rejecting physicalism and biologism while at the same time supporting the idea of a universal science in the form of a general systems theory? One would have to think that the former means von Bertalanffy bases his thinking on the idea of irreducibility, while the latter seems only possible on the basis of the postulate of continuity. And the question arises: in von Bertalanffy's view, what constitutes the continuity?

For an answer to the questions raised here, the conception of a hierarchically ordered reality is essential. This conception, which features specific system laws for every level of reality, puts von Bertalanffy in a position to reject reductionism. In general terms, one can say that in his view it is incorrect to explain "higher" system level phenomena with a theory developed for a "lower," that is, less complex level. One who does so is taking a reductionistic approach to reality. That is however not the case when one moves in the opposite direction, taking the path of expansionism instead of reductionism. Then one endeavors to "stretch" concepts and laws that bear upon a less complex level in such a way that what is more complex can thereby be grasped as well. This path of progressive expansion and generalization implies in principle the promise of a universal science, a general systems theory that would be valid for every level of the system hierarchy.

Thus the conception of a hierarchical architecture of reality with levels of increasing complexity puts von Bertalanffy in a position to reject reductionism in the sense of reducing the "higher" system level to the "lower" system level. At the same time, this conception offers the possibility of holding onto the deeper motive that underlies reductionism, namely, the desire to get a theoretical-scientific grasp of the whole of reality. To this end von Bertalanffy postulates a formal uniformity of reality, which he sometimes refers to as structural uniformity.[17] The unitary principle or common denominator to which reality is "reduced" is thus not found in a single system level but is given with the postulated structural uniformity that spans all the levels of the system hierarchy. This structural uniformity does not eliminate the autonomy of the different system levels. "When emphasizing general structural isomorphies of different levels, it asserts, at the same time, their autonomy and possession of specific laws."[18]

One may doubt whether there is a real difference between expansionism and reductionism. In fact, expansionism can be regarded as a kind of reductionism because it amounts to a reduction of reality to the formal-mathematical. That von Bertalanffy fails to recognize this is the consequence of his conception of reality, in which he distinguishes between a hierarchic order of systems on the one hand and a formal-mathematical dimension of reality on the other hand. Parallel with this distinction there are the special systems sciences concerning different kinds of systems and a general systems science of a formal-mathematical nature. However, one may argue that mathematics is not principally different from sciences such as physics, biology, sociology, etc. Mathematics is then concerned with its own irreducible realm of reality. We will elaborate on this point in the concluding considerations. Let us first analyze Simon's concept of hierarchy.

THE CONCEPT OF HIERARCHY IN SIMON

While von Bertalanffy as a biologist is brought into confrontation with the issue of reductionism or antireductionism via the bio-physical problem, Simon is compelled to face it in his field via the psychophysical problem. He has to deal with a modern variant of the controversy in the widely discussed question: Can a computer think?

For several decades advocates of so-called "strong AI" (Artificial Intelligence)—a term borrowed from Searle[19]—have asserted that it is just a matter of time before electronic computers are able to take the place of the human mind. Underlying this fundamental position is the idea that we are all part of a world that is controlled down to its most minute details by exact physical-chemical laws. Even our brains, which seem to direct all our activities, are controlled by the same exact laws. A picture emerges of the entire operation of the human brain as nothing more than the execution of an unusually complex algorithm. This mechanistic-reductionist hypothesis that the human mind is a machine that operates according to known natural laws has been challenged forcefully from various sides, recently by the British mathematician and physicist Roger Penrose.[20] Penrose supports to some extent the view that the operation of the brain is indeed largely algorithmic. However, he also believes that the question of reducing mind to machine has not been settled.

In the controversy touched upon here between those who believe human thought is reducible to regularities of a physical-mathematical nature and those who defend its irreducibility, Herbert Simon is a familiar spokesman for the former. Yet while he shows himself to be an indefatigable defender of mechanistic reductionism, Simon believes he can also contribute from his standpoint to bridging the opposition between the two schools. The controversy between reductionism and holism does not stem, so he believes, from a fundamental contradiction of a scientific character but is to be viewed rather as a *"cause de guerre* between scientists and humanists."[21] The holistic counterpole to reductionism arises, he maintains, from a humanistic concern for the consequences of reductionism for people and human dignity. Indeed, for complex systems such as human thought behaviors, for example, it is true that "the whole is more than the sum of its parts." Yet one ought not to construe this "more than" in a deeper metaphysical sense, but should take it just in the practical sense that, given the properties of the parts and the laws that control their coordination, it is no trivial matter to infer from them the properties of the whole. In short, so Simon says: "In the face of complexity an in-principle reductionist may be at the same time a pragmatic holist."[22]

Thus Simon recommends that we act as if holism is true while maintaining that human thought patterns and artifacts are subject to the same laws of nature as the natural world.[23] The conflicts that come to light in the debate about whether the computer can think consist, according to Simon, chiefly in the confrontation between two views of the nature of humanity. In the one view, the human person is set apart from the nature surrounding him or her, while in the other conception mankind is regarded as a part of nature. Since the rise of modern science, humans learned step by step that they are not unique beings. The view that rests on human apartness from the rest of nature has thereby lost its basis:

> With Copernicus and Galilei, he ceased to be the species located at the center of the universe, attended by sun and stars. With Darwin, he ceased to be the species created and specially endowed by God with soul and reason. With Freud, he ceased to be the species whose behavior was—potentially—governable by rational mind. As we begin to produce mechanisms that think and learn, he has ceased to be the species uniquely capable of complex, intelligent manipulation of his environment.[24]

The diminishing value of the position of mankind in the universe, Simon maintains, has not affected human dignity. He says:

> I am confident that man will, as he has in the past, find a new way of describing his place in the universe—a way that will satisfy his needs for dignity and for purpose. But it will be a way as different from the present one as was the Copernican from the Ptolemaic.[25]

And in words having a similar thrust he says elsewhere:

> We have learned that our human worth does not depend upon our remaining apart from nature. On the contrary, the human future depends on our being truly and fully a part of nature. Information processing psychology, far from threatening our place in things, gives us further evidence of the closeness of our kinship with all creation.[26]

Simon is undoubtedly correct that criticism of reductionism is very often inspired by fear of a reduced view of mankind. That is true in the case of von Bertalanffy, who eventually extended his critique of reductionism in the natural sciences to the human sciences with a sharp attack on the mechanization or robotization of the prevalent picture of mankind. In the twentieth century, von Bertalanffy writes, sociology and psychology have fallen increasingly under the spell of "...a positivistic-mechanistic-reductionistic approach which can be epitomized as the *robot model of man*."[27]

The contrast between von Bertalanffy and Simon appears clearly when the latter invokes the model of a robot to elucidate human thought from the perspective of systems theory. Human thought is, for Simon, the adaptive behavior of a system in a complex environment. The complexity of human thought as a behaving system must therefore be understood, according to Simon, in terms of the complexity of the environment. The environment, consisting of a certain problem, functions as a mold for human thought. His central thesis respecting the psychology of human thought is accordingly the following: *"A man, viewed as a behaving system, is quite simple. The apparent complexity of his behavior over time is largely a reflection of the complexity of the environment in which he finds himself."*[28]

According to Simon, complex systems, regardless of their kind, can be described as hierarchical structures. Complexity takes the form of hierarchy. By hierarchy Simon understands a constellation of mutually connected components or component systems each of which in its turn has a hierarchical structure, until one reaches the level of elementary subsystems. The word "hierarchy" is often used in a narrower meaning than Simon is giving here. What it refers to in that case is what Simon calls a "formal hierarchy." The latter is defined as a complex system in which each of the subsystems is subordinated by an authority relation. "More exactly,...each system consists of a 'boss' and a set of subordinate systems. Each of the subsystems has a 'boss' who is the immediate subordinate of the boss of the system."[29] Hierarchy in the broader sense means in fact a structure of parts-in-parts. Elaborating on Simon's ideas, Koestler called this the holon-character of systems: each system is a whole with respect to parts at a lower level and a part with respect to some whole at a higher level.[30]

That the complexity of reality takes the shape of a hierarchy is explained simply, according to Simon, from evolution. Simon uses the now-familiar parable of two watchmakers, Hora and Tempus, to explain that complex systems will evolve much more quickly from simple systems if there are stable intermediate forms. The watches Hora developed were no less complex than those of Tempus. Both men used about a thousand components for their products. Yet Hora was more successful than Tempus. "Tempus had so constructed his that if he had one partly assembled and had put it down—to answer the phone, say—it immediately fell to pieces and had to be reassembled from the elements." But Hora had so designed his watches

> that he could put together subassemblies of about ten elements each. Ten of these subassemblies, again, could be put together into a larger subassembly; and a system of ten of the latter subassemblies constituted

the whole watch. Hence, when Hora had to put down a partly assembled watch to answer the phone, he lost only a small part of his work, and he assembled his watches in only a fraction of the man-hours it took Tempus.[31]

CONCLUDING CONSIDERATIONS

A central question in our reflections here has been whether systems thinking, and in particular the systems-theoretical concept of hierarchy, can provide a key for overcoming reductionism in the sciences. Reductionism, so we argued, is implicit in the postulate of continuity and the desire to gain a theoretical grasp of reality in its unity and totality.

Of von Bertalanffy one can say that he is aware of the shortcomings and dangers of (mechanistic) reductionism. Yet the concept of hierarchy that he introduces is inadequate to overcome reductionism and the exaggerated expectations of theoretical thought underlying reductionism. The expansionism in science he advocates also leads to reduction of the diversity of reality to a single theoretical common denominator, namely, that of the logical-mathematical. The opposition between reductionism and expansionism disappears when reality is conceived as a hierarchical order starting at the lowest level with the mathematical domain and running upwards to the physical, the biological, etc.

With his view of a hierarchical order, von Bertalanffy correctly stresses the diversity in reality. Yet this diversity is still done insufficient justice as a result of his failure to take into account the valuable distinction between "entities" and "modes of existence" in the sense in which these were elaborated by the Dutch philosopher Dooyeweerd.[32] The stratification of increasing complexity in reality is regarded by von Bertalanffy as a hierarchy of systems or entities. In Dooyeweerd, by contrast, aside from the existence of systems with their qualifying characteristics, the diversity in reality is related to a hierarchical order of "modes of existence" or "modes of experience." These modes concern the rich diversity of aspects that we apprehend in things or systems. One can say that the existence of these modes of experience besides entities does not go entirely unnoticed in von Bertalanffy, particularly where he speaks of structural uniformities at the various system levels. From the point of view of irreducible modes of existence, von Bertalanffy's unification of the sciences on the basis of formal similarities (isomorphisms) appears to be a form of reduction of reality to the formal-mathematical. In this we see the opposition of two metaphysical beliefs about the nature of reality, namely the view of reality as discontinuous as opposed to the view of reality as continuous.

Where von Bertalanffy still endeavors to counter reductionism in the sciences, Simon adopts reductionism as his *a priori* starting point. That is, in his systems-theoretical interpretation of the complexity of reality ("The Architecture of Complexity"), the scheme of whole and parts is considered to be an all-determining context. The hierarchy of reality expressing an increasing complexity is explained in Simon entirely in terms of this scheme, while real modal or qualitative differences that present themselves to our experience are ignored. A difficulty that emerges, then, is that since everything that appears is "nature," it becomes impossible to identify what is typically human. Simon is apparently unaware that this difficulty leads to a remarkable tension in his thought. For while he needs the distinction between the natural and the artificial in order to develop "The Sciences of the Artificial," this point of departure is undermined when everything in reality is named "nature." In other words, although man is initially associated with the "artificial world," the final outcome of Simon's reflections is to see him as "fully a part of nature."

Our conclusion must be that current systems-theoretical conceptions of a hierarchical order of reality are inadequate for purposes of overcoming reductionism. It is interesting to notice, however, that in the circle of systems theorists an attempt has been made in recent years, by J.D.R. de Raadt,[33] to proceed in the line of the "multi-modal view" developed by Dooyeweerd. Further philosophical elaboration of this perspective in the field of systems thinking is a challenge for the future.

NOTES

1. Translated from the Dutch by Herbert Donald Morton.

2. This paper addresses primarily ontological reductionism. The resolution of all phenomena into physical events is one kind of ontological reductionism, namely physicalism. Ontological reduction may include theory reduction, but this is not necessarily the case. On the history and definition of the concept reductionism see J. Ritter and K. Gründer, *Historisches Wörterbuch der Philosophie* (Darmstadt: Wissenschaftliche Buchgesellschaft, 1992), band 8, 378-84.

3. E. Jantsch, *The Self-Organizing Universe: Scientific and Human Implications of the Emerging Paradigm of Evolution*, 5th ed. (Oxford and New York: Pergamon Press, 1989).

4. Jantsch, 242.

5. Jantsch, 23.

6. The literature on the concept of hierarchy in systems thinking includes, of course, much more than the contributions of these two writers. J.G. Miller presents a hierarchy of "living systems" in connection with a "general living systems theory" (*Living Systems* [New York: McGraw-Hill, 1979]). Hierarchies of systems comprehending the whole of reality may also be found in M. Bunge, "The Metaphysics, Epistemology and Methodology of Levels," in *Hierarchical Structures*, edited by L.L. Whyte, A.G. Wilson and D. Wilson (New York: American Elsevier, 1969), 17-28; and E. Laszlo, *Introduction to Systems Philosophy: Toward a New Paradigm of Contemporary Thought* (New York: Harper and Row, 1972). The former calls his view "integrated pluralism," which is to say "an ontology that proclaims both the diversity and the unity of the world." A historical introduction of the concept of hierarchy is provided by L.L. Whyte in a collection which he, A.G. Wilson and D. Wilson edited (L.L. Whyte, "Structural Hierarchies: A Challenging Class of Physical and Biological Problems," *Hierarchical Structures*, edited by L.L. Whyte, A.G. Wilson and D. Wilson [New York: American Elsevier, 1969], 3-16). For more recent literature exploring the concept of hierarchy, see S.N. Salthe, *Evolving Hierarchical Systems: Their Structure and Representation* (New York: Columbia University Press, 1985); S.N. Salthe, "Self-Organization of/in Hierarchically Structured Systems," *Systems Research* 6 (1989): 199-208; U. Zylstra, "Living Things as Hierarchically Organized Structures," *Synthese* 91 (1992): 111-33.

7. Various kinds of hierarchies can be distinguished in systems thinking. This paper takes the hierarchy concept in a broad sense, namely as a structure of parts-in-parts (see note 26). Whether or not selection of a particular concept of hierarchy is influenced by certain metaphysical beliefs will not be investigated.

8. For a critical discussion of the concept of the open system, see Strijbos, "The Concept of the "Open System"—Another Machine Metaphor for the Organism?" in *Facets of Faith and Science. Volume 3: The Role of Beliefs in the Natural Sciences*, edited by J.M. van der Meer (Lanham: The Pascal Centre for Advanced Studies in Faith and Science/University Press of America, 1996).

9. L. von Bertalanffy, *Problems of Life: An Evaluation of Modern Biological Thought* (London: Watts and Company, 1953), 23.

10. L. von Bertalanffy, *General System Theory: Foundations, Development, Applications* (Harmondsworth: Penguin Books, 1968), 27.

11. Von Bertalanffy, *General System Theory*, 87.

12. Von Bertalanffy, *General System Theory*, 25; italics added.

13. Von Bertalanffy, *General System Theory*, xix.

14. Although vitalism opposes mechanism it can be doubted whether it really escapes a mechanistic framework. In fact it seems to accept the machine-view of the organism, while supplementing it with the notion of a vital force.

15. Von Bertalanffy, *General System Theory*, 88.

16. Von Bertalanffy, *General System Theory*, 88.

17. See von Bertalanffy, *General System Theory*, 32, 48, 63, 87.

18. Von Bertalanffy, *General System Theory*, 88.

19. J.R. Searle, "Minds, Brains and Programs," *Behavioral and Brain Sciences* 3 (1980): 417-24.

20. R. Penrose, *The Emperor's New Mind* (Oxford: Oxford University Press 1989).

21. H.A. Simon, *The Sciences of the Artificial* (Cambridge and London: MIT Press, 1969; revised 1982), 195 n.4.

22. Simon, *Sciences of the Artificial*, 195.

23. See Simon, *Sciences of the Artificial*, chap. 1.

24. H.A. Simon, *The New Science of Management Decision* (Englewood Cliffs: Prentice Hall, 1960; revised 1977), 37.

25. Simon, *Management Decision*, 37.

26. H.A. Simon, *Man and His Tools: Technology and the Human Condition*, Duijker Lecture 1981 (Amsterdam: Intermediair Bibliotheek, 1981), 16.

27. L. von Bertalanffy, *Robots, Men and Minds: Psychology in the Modern World* (New York: George Braziller, 1967), 7.

28. Simon, *Sciences of the Artificial*, 65.

29. See Simon, *Sciences of the Artificial*, chap. 7; this chapter has been published in 1962 as an article.

30. See in A. Koestler, *The Ghost in the Machine* ([London: Hutchison, 1967]; reprint of 1967 ed. [New York: Arkana Books, 1989]), chap. 3; and A. Koestler, "The Tree and the Candle," in *Unity through Diversity: Festschrift for Ludwig von Bertalanffy*, edited by W. Gray and N.D. Rizzo (New York, London, Paris: Gordon and Breach Science Publishers, 1971), 287-315. What Simon calls a "formal hierarchy" is a "control hierarchy" in Koestler's terminology. There have been several attempts to classify hierarchies into categories. Following the distinction between structure and function one can distinguish between "structural" hierarchies, which emphasize the spatial aspect (anatomy, topology) of a system, and "functional" hierarchies, which

emphasize process in time. A common feature of all hierarchies is that they have a "part within part" character (Koestler, *Ghost in the Machine*, 59).

31. Simon, *Sciences of the Artificial*, 200.

32. H. Dooyeweerd, *A New Critique of Theoretical Thought*, 4 vols. (Amsterdam: H.J. Paris Publishers; Philadelphia: Presbyterian and Reformed Publishing Company, 1953-1958). See also R.A. Clouser, *The Myth of Religious Neutrality: An Essay on the Hidden Role of Religious Belief in Theories* (Notre Dame: University of Notre Dame Press, 1991).

33. J.D.R. de Raadt, "Multi-Modal Systems Design: A Concern for the Issues that Matter," *Systems Research* 6 (1989): 17-25; J.D.R. de Raadt, *Information and Managerial Wisdom* (Pocatello: Paradigm Publications, 1991).

Bibliography

Alcinous. *Didaskalikos*. In *Platonis Dialogi secundum thrasylli tetralogias dispositi*, edited by C.F. Hermann. 6 vols. in 3. Leipzig: Teubner, 1859.

Alexander, H.G. *The Leibniz-Clarke Correspondence*. Manchester: University of Manchester Press, 1956.

Alexander, R. *Optima for Animals*. London: Edward Arnold, 1982.

Andrews, E.H. "The Age of the Earth." In *Creation and Evolution: When Christians Disagree*, edited by O.R. Barclay, 64. Downers Grove: InterVarsity Press, 1985.

Anonymous. "Mr. Proctor's New Work." *Astronomical Register* 8 (1870): 143-144.

Anonymous. "Other Worlds than Ours." *English Mechanic* 11 (1870): 271.

Anonymous. "Other Worlds than Ours." *Quarterly Journal of Science* 7 (1870): 367-373.

Arp, H.C. *Quasars, Redshifts and Controversies*. Berkeley: Interstellar Media, 1987.

Attwater, R. *Adam Schall: A Jesuit at the Court of China*. Adapted from the French of Joseph Duhr, S.J. (*Un Jésuite en Chine, Adam Schall*. Paris, 1936). London: Geoffrey Chapman, 1963.

Augustine, T. *Confessions*. New York: Penguin Classics, 1961.

Austin, W.H. "Isaac Newton on Science and Religion." *Journal of the History of Ideas* 31 (1970): 521-542.

Ayala, F. "Evolution, The Theory of." *Encyclopedia Britannica*. Fifteenth Edition. Chicago: Encyclopedia Britannica, 1988.

Ayala, F., and J. Valentine. *Evolving: The Theory and Processes of Organic Evolution*. Menlo Park: Benjamin/Cummings, 1979.

Beardsley, T. "Smart Genes." *Scientific American* 265 (1991): 86-95.

Bechtel, W., and R.C. Richardson. *Discovering Complexity: Decomposition and Localization Strategies in Scientific Research.* Princeton: Princeton University Press, 1993.

Beck, S. "Natural Science and Creationist Theology." *BioScience* 32 (1982): 738-742.

Bell, J.S. "On the Problem of Hidden Variables in Quantum Mechanics." *Reviews of Modern Physics* 38 (1966): 447-452.

Bernard, H. *Matteo Ricci's Scientific Contribution to China.* Translated by E.C. Werner. Peiping: Henri Vetch, 1935.

Bettray, J. *Die Akkommodationsmethode des P. Matteo Ricci S.I. in China. Analecta Gregoriana* 76. Series Facultatis Missiologicae Sectio B (no. 1). Rome: Gregorian University, 1955.

Bienvenue, R.T., and M. Feingold, eds. *In the Presence of the Past.* Dordrecht: Kluwer Academic Publishers, 1991.

Birtel, F.T., ed. *Religion, Science and Public Policy.* New York: Crossroad Publishing Company, 1987.

Bohm, D. "A Suggested Interpretation of the Quantum Theory in Terms of 'Hidden' Variables. I and II." *Physical Review* 85 (1952): 166-193.

Bohr, N. *Atomic Physics and Human Knowledge.* New York: John Wiley, 1958.

Bohr, N. *Essays, 1958-1962, on Atomic Physics and Human Knowledge.* New York: Interscience, 1963.

Bolter, J.D. *Turing's Man: Western Culture in the Computer Age.* Chapel Hill: University of North Carolina Press, 1984.

Born, M. *Einstein's Theory of Relativity.* New York: Dover Publications, 1962.

Borstnik, B., D. Pumpernik, and G.L. Hofacker. "Point Mutations as an Optimal Search Process in Biological Evolution." *Journal of Theoretical Biology* 125 (1987): 249-268.

Bosmans, H. "Ferdinand Verbiest: Directeur de l'Observatoire de Peking (1623-1688)." *Revue des questions scientifiques* 21: 195-273. Louvain: Société scientifique de Bruxelles, 1912.

Bowler, P. *Evolution: The History of an Idea.* Berkeley: University of California Press, 1983.

Boyd, C.A.R., and D. Noble, eds. *The Logic of Life: The Challenge of Integrative Physiology.* Oxford: Oxford University Press, 1993.

Boyle, R. *The Works of the Honourable Robert Boyle*, edited by T. Birch. 6 vols. London: Millar, 1772.

Brady, R.H. "On the Independence of Systematics." *Cladistics* 1 (1985): 113-126.

Brenner, S., W. Dove, I. Herskowitz, and R. Thomas. "Genes and Development: Molecular and Logical Themes." *Genetics* 126 (1990): 479-486.

Brooke, J.H. "Natural Theology and the Plurality of Worlds." *Annals of Science* 34 (1977): 221-286.

Brooke, J.H. *Newton and the Mechanistic Universe.* Milton Keynes: Open University Press, 1974.

Brooke, J.H. "The Relations between Darwin's Science and His Religion." In *Darwinism and Divinity: Essays on Evolution and Religious Beliefs*, edited by J. Durant, 40-75. London: Basil Blackwell, 1985.

Brooke, J.H. *Science and Religion: Some Historical Perspectives.* Cambridge: Cambridge University Press, 1991.

Brooks, D.R., J. Collier, B.A. Maurer, J.D.H. Smith, and E.O. Wiley. "Entropy and Information in Evolving Biological Systems." *Biology and Philosophy* 4 (1989): 407-432.

Brooks, R.S. *The Relationships between Natural Philosophy, Natural Theology and Revealed Religion in the Thought of Newton and Their Historiographic Relevance.* Unpublished Ph.D. dissertation from Northwestern University, Evanston, Illinois, 1976.

Bryant, P.J., and P. Simpson. "Intrinsic and Extrinsic Control of Growth in Developing Organs." *Quarterly Review of Biology* 59 (1984): 387-415.

Buccholtz, K-D. *Isaac Newton als Theologe; ein Beitrag zum Gespräch zwischen Naturwissenschaft und Theologie.* Witten: Luther-Verlag, 1965.

Bunge, M. "A Ghost-Free Axiomatization of Quantum Mechanics." In *Quantum Theory and Reality*, edited by M. Bunge, 105-117. New York: Springer-Verlag, 1967.

Bunge, M. "The Metaphysics, Epistemology and Methodology of Levels." In *Hierarchical Structures*, edited by L.L. Whyte, A.G. Wilson and D. Wilson, 17-28. New York: American Elsevier, 1969.

Bunge, M. *Scientific Materialism*. Dordrecht: D. Reidel Publishing Company, 1981.

Burian, R. "Why the Panda Provides No Comfort to the Creationist." *Philosophica* 37 (1986): 11-26.

Burkert, W. *Lore and Science in Ancient Pythagoreanism*. Translated by E.L. Minar. Cambridge: Harvard University Press, 1972.

Burns, E.M., R.E. Lerner, and S. Meacham. *Western Civilizations: Their History and Their Culture*. Tenth Edition. New York: W.W. Norton and Company, 1984.

Cain, A.J. "The Perfection of Animals." In *Viewpoints in Biology*. Volume 3, edited by J.D. Carthy and C.L. Duddington, 36-63. London: Butterworths, 1964.

Cairns-Smith, A.G. *The Life Puzzle*. Edinburgh: Oliver and Boyd, 1971.

Caron, F. "Eucaryotic Codes." *Experientia* 46 (1990): 1106-1117.

Cartmill, M. *A View to a Death in the Morning: Hunting and Nature through History*. Cambridge: Harvard University Press, 1993.

Castillejo, D. *The Expanding Force in Newton's Cosmos*. Madrid: Ediciones de Arte y Bibliofilia, 1981.

Chiang, F.T. "The History of Matteo Ricci's Missionary to China and the Meaning of His Book *The True Idea of God*." In *International Symposium on Chinese-Western Cultural Interchange in Commemoration of the 400th Anniversary of the Arrival of Matteo Ricci, S.J. in China*. Taiwan, Republic of China, 1983.

Churchland, P. and C. Hooker, eds. *Images of Science*. Chicago: University of Chicago Press, 1985.

Clarke, S. *A Collection of Papers, Which Passed Between the Late Learned Mr. Leibnitz, and Dr. Clarke, in the Years 1715 and 1716*. London: N.p., 1717.

Clavelin, M. *The Natural Philosophy of Galileo: Essay on the Origins and Formation of Classical Mechanics*. Translated by A.J. Pomerans. Cambridge: MIT Press, 1974.

Clement of Alexandria. Translated by G.W. Butterworth. Reprint of 1919 Edition, Loeb Classical Library. Cambridge: Harvard University Press, 1982.

Clouser, R.A. *The Myth of Religious Neutrality: An Essay on the Hidden Role of Religious Belief in Theories*. Notre Dame: University of Notre Dame Press, 1991.

Clouser, R.A. "A Sketch of Dooyeweerd's Philosophy of Science." In *Facets of Faith and Science. Volume 2: The Role of Beliefs in Mathematics and the Natural Sciences: An Augustinian Perspective* edited by J.M. van der Meer. Lanham: The Pascal Centre for Advanced Studies in Faith and Science/University Press of America, 1996.

Cobb, J.B. *A Christian Natural Theology: Based on the Thought of Alfred North Whitehead*. Philadelphia: Westminster Press, 1965.

Cornell, J. "God's Magnificent Law: The Bad Influence of Theistic Metaphysics on Darwin's Estimation of Natural Selection." *Journal of the History of Biology* 20 (1987): 381-412.

Courtenay, W.J. "Nominalism and Late Medieval Religion." In *The Pursuit of Holiness in Late Medieval and Renaissance Religion*, edited by C. Trinkaus and H. Oberman, 26-59. Leiden: E.J. Brill, 1974.

Craig, W.L. "Philosophical and Scientific Pointers to Creation *ex nihilo*." *Journal of the American Scientific Affiliation* 32 (1980): 5-13.

Cramer, J.A. "Science, Scientism and Christianity: The Ideas of D.M. MacKay." *Journal of the American Scientific Affiliation* 37 (1985): 142-148.

Crombie, A.C. "The Sources of Galileo's Early Natural Philosophy." In *Reason, Experiment, and Mysticism in the Scientific Revolution*, edited by M.L. Righini-Bonelli and W.R. Shea, 157-175. New York: Science History Publications, 1975.

Crowe, M.J. *The Extraterrestrial Life Debate 1750-1900: The Idea of a Plurality of Worlds from Kant to Lowell*. Cambridge: Cambridge University Press, 1986.

Crowe, M.J. "Richard Proctor and Nineteenth-Century Astronomy." *History of Science Society Meeting* 1 (1989).

Cummins, J.S. *A Question of Rites: Friar Domingo Navarrete and the Jesuits in China*. Aldershot: Scolar Press, 1993.

Dallmayr, F. *Between Freiburg and Frankfurt*. Amherst: University of Massachusetts Press, 1991.

Damon, P.E., D.J. Donahue, B.H. Gore, A.L. Hatheway, A.J.T. Jull, T.W. Linick, P.J. Sercel, L.J. Toolin, C.R. Bronk, E.T. Hall, R.E.M. Hedges, R. Housley, I.A. Law, C. Perry, G. Bonani, S. Trumbore, W. Woelfli, J.C. Ambers, S.G.E. Bowman, M.N. Leese, and M.S. Tite. "Radiocarbon Dating of the Shroud of Turin." *Nature* 337 (1989): 611-615.

Darnbrough, C. "Genes—Created but Evolving." In *Concepts in Creationism*, edited by E.H. Andrews, W. Gitt, and W.J. Ouweneel, 241-266. Herts: Evangelical Press, 1986.

Darwin, C. *On the Origin of Species*. Facsimile Reprint of First Edition. Cambridge: Harvard University Press, 1964.

Darwin, C. *The Various Contrivances by which Orchids are Fertilized by Insects*. Chicago: University of Chicago Press, 1984.

Davies, P. *God and the New Physics*. New York: Simon and Schuster, 1983.

Davis, B.D. "Molecular Genetics and the Foundations of Evolution." *Perspectives in Biology and Medicine* 28 (1985): 251-268.

Davis, E.B. "Blessed are the Peacemakers: Rewriting the History of Christianity and Science." *Perspectives on Science and Christian Faith* 40 (1988): 47-52.

Davis, E.B. "Book Review of *Creation, Nature, and Political Order in the Philosophy of Michael Foster (1903-1959): The Classic "Mind" Articles and Others, with Modern Critical Essays*, edited by C. Wybrow." *Isis* 85 (1994), 127-129. Reprinted with editorial changes and additions as "Christianity and Early Modern Science: Beyond War and Peace?" *Perspectives on Science and Christian Faith* 46 (1994): 133-135.

Davis, E.B. "Creation, Contingency, and Early Modern Science: The Impact of Voluntaristic Theology on Seventeenth-Century Natural Philosophy." Unpublished Ph.D. dissertation from the University of Indiana, 1984.

Davis, E.B. "God, Man, and Nature: The Problem of Creation in Cartesian Thought." *Scottish Journal of Theology* 44 (1991): 325-348.

Davis, E.B. "Newton's Rejection of the 'Newtonian World View.'" *Fides et Historia* 22 (1990): 6-20. Reprinted in *Science and Christian Belief* 3 (1991): 103-117. Also reprinted with minor additions as "Newton's Rejection of the 'Newtonian Worldview.': The Role of Divine Will in Newton's Natural Philosophy." in *Facets of Faith and Science. Volume 3: The Role of Beliefs in the Natural Sciences*, edited by J.M. van der Meer. Lanham: The Pascal Centre for Advanced Studies in Faith and Science/University Press of America, 1996.

Dawkins, R. *The Blind Watchmaker: Why the Evidence of Evolution Reveals a Universe without Design.* New York: W.W. Norton and Company, 1986.

Deason, G.B. "Reformation Theology and the Mechanistic Conception of Nature." In *God and Nature: Historical Essays on the Encounter between Christianity and Science,* edited by D.C. Lindberg and R.L. Numbers, 167-191. Berkeley: University of California Press, 1986.

De Chardin, P.T. *The Phenomenon of Man.* London: Collins and New York: Harper and Row Publishers, 1959.

D'Elia, P.M. *Galileo in China: Relations through the Roman College between Galileo and the Jesuit Scientist-Missionaries (1610-1640).* Translated by R. Suter and M. Sciascia. Cambridge: Harvard University Press, 1960.

Demant, V.A. Memorial Sermon for M.B. Foster. *Christian Scholar* 43 (1960): 3-7.

de Raadt, J.D.R. *Information and Managerial Wisdom.* Pocatello: Paradigm Publications, 1991.

de Raadt, J.D.R. "Multi-Modal Systems Design: A Concern for the Issues that Matter." *Systems Research* 6 (1989): 17-25.

Descartes, R. *Le Monde, ou Traité de la lumière.* Translated by M.S. Mahoney. New York: Abaris Books, 1979

Descartes, R. *Oeuvres de Descartes,* edited by C. Adam and P. Tannery. 12 vols. Paris: Leopold Cerf. 1897-1910.

Desmond, A. *The Politics of Evolution.* Chicago: University of Chicago Press, 1989.

Dewan, E.M. "Consciousness as an Emergent Causal Agent in the Context of Control System Theory." In *Consciousness and the Brain: A Scientific and Philosophical Inquiry,* edited by G.G. Globus, G. Maxwell, and I. Savodnik, 181-198. New York: Plenum Press, 1976.

Dingle, H. "Philosophical Aspects of Cosmology." *Vistas in Astronomy* 1 (1960): 162-166.

Dobbs, B.J.T. *The Foundations of Newton's Alchemy: or, "The Hunting of the Greene Lyon."* Cambridge: Cambridge University Press, 1975.

Dobbs, B.J.T. *The Janus Faces of Genius: The Role of Alchemy in Newton's Thought.* Cambridge: Cambridge University Press, 1991.

Dobbs, B.J.T. "Newton's Alchemy and His Theory of Matter." *Isis* 73 (1982): 511-528.

Dobzhansky, T. "Nothing in Biology Makes Sense Except in the Light of Evolution." *American Biology Teacher* 35 (1973): 125-129.

Dobzhansky, T., F. Ayala, G.L. Stebbins, and J. Valentine. *Evolution*. San Francisco: W.H. Freeman, 1977.

Doolittle, R.R. "New Perspectives on Evolution Provided by Protein Sequences." In *New Perspectives on Evolution*, edited by L. Warren and H. Koprowski, 165-173. The Wistar Symposium Series, Volume 4. New York: John Wiley-Alan Liss, 1991.

Dooyeweerd, H. *A New Critique of Theoretical Thought*. 4 vols. Amsterdam: H.J. Paris Publishers and Philadelphia: Presbyterian and Reformed Publishing Company, 1953-1958.

Draper, J.W. History of the Conflict between Religion and Science. London: n.p, 1875.

du Halde, J.B. *Description géographique, historique, chronologique et physique de l'Empire de la Chine et de la Tartarie Chinoise*. 4 vols. La Haye: Henri Scheurleer, 1736.

Duhem, P. *The Aim and Structure of Physical Theory*. Translated by P.P. Wiener. Princeton: Princeton University Press, 1906/1954.

Dunne, G.H. *Generation of Giants: The Story of the Jesuits in China in the Last Decades of the Ming Dynasty*. Notre Dame: University of Notre Dame Press, 1962.

Dyson, F.J. *Infinite in All Directions: An Exploration of Science and Belief*. New York: Harper and Row Publishers, 1988.

Eldredge, N., and J. Cracraft. *Phylogenetic Patterns and the Evolutionary Process*. New York: Columbia University Press, 1980.

Ellis, G.F.R. "Alternatives to the Big Bang." *Annual Review of Astronomy and Astrophysics* 22 (1984): 157-184.

Ellis, G.F.R. "Cosmology and Verifiability." *Quarterly Journal of the Royal Astronomical Society* 16 (1975): 245-264.

Ellis, G.F.R. "Is the Universe Expanding?" *General Relativity and Gravitation* 9 (1978): 87-94.

Evans, C.S. *Preserving the Person: A Look at the Human Sciences.* Downers Grove: InterVarsity Press, 1977. Reprint. Grand Rapids: Baker Book House, 1982.

Fine, A. "The Natural Ontological Attitude." In *Scientific Realism*, edited by J. Leplin, 83-108. Berkeley: University of California Press, 1984.

Ford, K.W. *The World of Elementary Particles.* New York: Blaisdell, 1965.

Foster, J. *The Case for Idealism.* London: Routledge and Kegan Paul, 1982.

Foster, M.B. "The Christian Doctrine of Creation and the Rise of Modern Natural Science." *Mind* 43 (1934): 446-468.

Foster, M.B. "Christian Theology and Modern Science of Nature." *Mind* 44 (1935): 439-66 and 45 (1936): 1-27.

Fox, T.D. "Natural Variation in the Genetic Code." *Annual Review of Genetics* 21 (1987): 67-91.

Frair, W., and P. Davis. *A Case for Creation.* Third Edition. Chicago: Moody Press, 1983.

Futuyma, D. "Evolution as Fact and Theory." *Bios* 56 (1985): 3-13.

Futuyma, D. *Evolutionary Biology.* Sunderland: Sinauer Associates, 1979. Second Edition, 1986.

Futuyma, D. *Science on Trial: The Case for Creation.* New York: Pantheon Books, 1983.

Galilei, G. *The Assayer.* In *The Controversy on the Comets of 1618*, edited by S. Drake and C.D. O'Malley. Philadelphia: University of Pennsylvania Press, 1960.

Galilei, G. *Dialogue Concerning the Two Chief World Systems—Ptolemaic and Copernican.* Translated by S. Drake. Berkeley: University of California Press, 1953.

Galilei, G. *Discourses on Two New Sciences.* Translated by S. Drake. Madison: University of Wisconsin Press, 1974.

Galilei, G. *Letter to the Grand Duchess Christina.* In *Discoveries and Opinions of Galileo*, edited by S. Drake, 175-216. Garden City: Doubleday, 1957.

Garen, A. "Looking for the Homunculus in Drosophila." *Genetics* 131 (1992): 5-7.

266 Facets of Faith and Science

Gernet, J. China and the Christian Impact: A Conflict of Cultures. Translated by J. Lloyd. Cambridge: Cambridge University Press, 1985. Translation of Chine et christianisme. Paris: Éditions Gallimard, 1982.

Ghiselin, M. The Triumph of the Darwinian Method. Second Edition. Chicago: University of Chicago Press, 1984.

Gilkey, L. Creationism on Trial: Evolution and God at Little Rock. Minneapolis: Winston Press, 1985.

Gillespie, N. Charles Darwin and the Problem of Creation. Chicago: University of Chicago Press, 1979.

Gitt, W. "Information: The Third Fundamental Quality." Siemens Review 56, no. 6 (1989): 2-7.

Gould, S.J. "Darwinism Defined: The Difference between Fact and Theory." Discover 8 (January, 1987): 64-70.

Gould, S.J. Ever Since Darwin: Reflections on Natural History. New York: W.W. Norton and Company, 1977.

Gould, S.J. "Evolution and the Triumph of Homology, or Why History Matters." American Scientist 74 (1986): 60-69.

Gould, S.J. "Evolution as Fact and Theory." Discover 2 (1981): 34-37.

Gould, S.J. Hen's Teeth and Horse's Toes: Further Reflections on Natural History. New York: W.W. Norton and Company, 1983.

Gould, S.J. The Panda's Thumb. New York: W.W. Norton and Company, 1980.

Gould, S.J. "The Panda's Thumb of Technology." In Bully for Brontosaurus, 59-75. New York: W.W. Norton and Company, 1991.

Gould, S.J. Wonderful Life: The Burgess Shale and the Nature of History. New York: W.W. Norton and Company, 1989.

Greer, T.H. A Brief History of the Western World. Fourth Edition. New York: Harcourt Brace Jovanovich, 1982

Gribbon, J. "'Bunched' Redshifts Question Cosmology." New Scientist 132 (1991): 1800-1801.

Gruner, R. "Science, Nature, and Christianity." Journal of Theological Studies 26 (1975): 55-81. Reprinted in Creation, Nature, and Political Order in the

Philosophy of Michael Foster (1903-1959): The Classic "Mind" Articles and Others, with Modern Critical Essays, edited by C. Wybrow, 213-243. Lewiston and Lampeter, Wales: Edwin Mellen Press, 1992.

Haas, J.W., Jr. "Complementarity and Christian Thought—An Assessment: 2. Logical Complementarity." *Journal of the American Scientific Affiliation* 35 (1983): 203-209.

Haeckel, E. *The History of Creation.* New York: D. Appleton, 1876.

Hartshorne, C. *The Logic of Perfection.* LaSalle: Open Court Publishing Company, 1962.

Hasker, W. "MacKay on Being a Responsible Mechanism: Freedom in a Clockwork Universe." *Christian Scholar's Review* 8 (1978): 130-140.

Hawking, S.W. *A Brief History of Time.* London: Bantam Books, 1988.

Hedges, R.E.M. "Reply." *Nature* 337 (1989): 594.

Heisenberg, W. *The Physical Principles of the Quantum Theory.* New York: University of Chicago Press, 1930. Reprint. New York: Dover Press, 1949.

Hempel, C.G. *Philosophy of Natural Science.* Englewood Cliffs: Prentice Hall, 1966.

Henderson, J.B. *The Development and Decline of Chinese Cosmology.* New York: Columbia University Press, 1984.

Henry, J. "Henry More Versus Robert Boyle: The Spirit of Nature and the Nature of Providence." In *Henry More (1614-1687)*, edited by S. Hutton, 55-76. Dordrecht: Kluwer Academic Publishers, 1990.

Hesiod. *Hesiod: The Homerica Hymns and Homerica*, edited and translated by H.E. White. Reprint of the 1914 Edition, Loeb Classical Library. New York: G.P. Putnam's Sons, 1982.

Hick, J. "Evil, The Problem of." In *The Encyclopedia of Philosophy*, edited by P. Edwards. New York: Macmillan Publishers, 1967.

Hick, J. *An Interpretation of Religion.* New Haven: Yale University Press, 1989.

Hodge, J. "The Development of Darwin's General Biological Theorizing." In *Evolution from Molecules to Man*, edited by D.S. Bendall, 43-62. Cambridge: Cambridge University Press, 1983.

Hoffman, A. *Arguments on Evolution: A Paleontologist's Perspective*. Oxford: Oxford University Press, 1989.

Homer. *The Odyssey*, edited and translated by A.T. Murray. 2 vols. Loeb Classical Library. New York: G.P. Putnam's Sons, 1919.

Hooykaas, R. *Religion and the Rise of Modern Science*. First American Edition. Grand Rapids: William B. Eerdmans Publishing Company, 1972.

Hooykaas, R. "Science and Theology in the Middle Ages." *Free University Quarterly* 3 (1954): 77-163.

Hoyle, F. *Astronomy and Cosmology*. San Francisco: Freeman, 1975.

Hoyle, F. *The Intelligent Universe*. New York: Holt, Rinehart and Winston, 1984.

Hull, D.L. "Darwin and the Nature of Science." In *Evolution from Molecules to Men*, edited by D.S. Bendall, 63-80. Cambridge: Cambridge University Press, 1983.

Hull, D.L. "A Function for Actual Examples in the Philosophy of Science." In *What the Philosophy of Biology Is*, edited by M. Ruse, 309-321. Dordrecht: Kluwer Academic Publishers, 1989.

Hutchinson, K. "Supernaturalism and the Mechanical Philosophy." *History of Science* 21 (1983): 297-333.

International Symposium on Chinese-Western Cultural Interchange in Commemoration of the 400th Anniversary of the Arrival of Matteo Ricci, S.J. in China. Taiwan, Republic of China, 1983.

Jacob, F. *The Possible and the Actual*. New York: Pantheon Books, 1982.

Jantsch, E. *The Self-Organizing Universe: Scientific and Human Implications of the Emerging Paradigm of Evolution*. Fifth Edition. Oxford and New York: Pergamon Press, 1989.

Johnson, P.E. *Darwin on Trial*. Washington: Regnery-Gateway, 1991.

Johnson, P.E. *Reason in the Balance: The Case Against Naturalism in Science*. Downer's Grove: InterVarsity Press, 1995

Jonas, H. *The Phenomenon of Life*. Chicago: University of Chicago Press, 1966.

Jukes, T. "Genetic code 1990. Outlook." *Experientia* 46 (1990): 1149-1157.

Jukes, T., and S. Osawa. "Recent Evidence for Evolution of the Genetic Code." In *Evolution of Life: Fossils, Molecules and Culture*, edited by S. Osawa and T. Honjo, 79-95. New York: Springer-Verlag, 1991.

Junker, R. *Rudimentäre Organe und Atavismen: Konstruktionsfehler des Lebens?* Berlin: Studium Integrale/Zeitjournal Verlag, 1989.

Junker, R., and S. Scherer. *Entstehung und Geschichte der Lebewesen.* Giessen: Weyel Lehrmittelverlag, 1988.

Kagan, D., S. Ozment, and F.M. Turner. *The Western Heritage.* Second Edition. New York: Macmillan Publishers, 1983.

Kant, I. *Critique of Judgment.* Translated with an Introduction by J.H. Bernard. New York: Hafner Press, 1951.

Kauffman, S.A. "Antichaos and Adaptation." *Scientific American* 265 (1991): 78-84.

Kelly, B. "Turin Shroud." *New Scientist* 119 (1988): 94.

Kenny, A. *Descartes: Philosophical Letters.* Oxford: Oxford University Press, 1970.

Kenyon, C., and B. Wang. "A Cluster of *Antennapedia*-Class Homeobox Genes in a Nonsegmented Animal." *Science* 253, no. 2 (1991): 516-517.

Kitcher, P. *Abusing Science: The Case Against Creationism.* Cambridge: MIT Press, 1982.

Kitcher, P. "Darwin's Achievement." In *Reason and Rationality in Natural Science*, edited by N. Rescher, 127-189. Lanham: University Press of America, 1985.

Klaaren, E.M. *Religious Origins of Modern Science. Belief in Creation in Seventeenth-Century Thought.* Grand Rapids: William B. Eerdmans Publishing Company, 1977.

Koestler, A. *The Ghost in the Machine.* London: Hutchison, 1967. Imprint of 1967 Edition. New York: Arkana Books, 1989.

Koestler, A. "The Tree and the Candle." In *Unity through Diversity: A Festschrift for Ludwig von Bertalanffy*, edited by W. Gray and N.D. Rizzo, 287-315. 2 vols. New York, London, Paris: Gordon and Breach Science Publishers, 1971.

Kohn, D. "Darwin's Ambiguity: The Secularization of Biological Meaning." *British Journal for the History of Science* 22 (1989): 215-239.

Kolakowski, L. *Religion.* Oxford: Oxford University Press, 1982.

Koyré, A. *From the Closed World to the Infinite Universe.* Baltimore: Johns Hopkins University Press, 1957.

Koyré, A. "Gravity an Essential Property of Matter?" In *Newtonian Studies,* edited by A. Koyré, 149-163. Cambridge: Harvard University Press, 1965.

Koyré, A. "Newton and Descartes." In *Newtonian Studies,* edited by A. Koyré, 53-114. Cambridge: Harvard University Press, 1965.

Koyré, A., and I.B. Cohen. "The Case of the Missing *Tanquam*: Leibniz, Newton and Clarke." *Isis* 52 (1961): 555-566.

Koyré, A., and I.B. Cohen. "Newton and the Leibniz-Clarke Correspondence with Notes on Newton, Conti, and Des Maizeau." *Archives internationales d'histoire des sciences* 15 (1962): 63-126.

Kubrin, D. "Newton and the Cyclical Cosmos: Providence and the Mechanical Philosophy." *Journal of the History of Ideas* 28 (1967): 325-346.

Kwok, D.W.Y. "*Ho* and *T'ung.*" In *Cosmology, Ontology, and Human Efficacy: Essays in Chinese Thought,* edited by R.J. Smith and D.W.Y. Kwok, 1-9. Honolulu: University of Hawaii Press, 1993.

Lakatos, I. *The Methodology of Research Programmes.* Cambridge: Cambridge University Press, 1980.

Landsberg, P.T., and D.A. Evans. "Free Will in a Mechanistic Universe?" *British Journal for the Philosophy of Science* 2 (1970): 343-358.

Laszlo, E. *Introduction to Systems Philosophy: Toward a New Paradigm of Contemporary Thought.* New York: Harper and Row Publishers, 1972.

Lattis, J. *Between Copernicus and Galileo: Christoph Clavius and the Collapse of Ptolemaic Cosmology.* Chicago: University of Chicago Press, 1994.

Laudan, L. "A Confutation of Convergent Realism." *Philosophy of Science* 48 (1981): 19-48.

Lauder, G. V. "Functional Morphology and Systematics: Studying Functional Patterns in a Historical Context." *Annual Review of Ecology and Systematics* 21.(1990): 317-40.

LaViolette, P.A. "Is the Universe Really Expanding?" *Astrophysical Journal* 301 (1984): 544-553.

[Leibniz, G.W.] *The Preface to Leibniz'* Novissima Sinica: *Commentary, Translation, Text by Donald F. Lach.* Honolulu: University of Hawaii Press, 1957.

Leibniz, G.W. *Theodicy.* LaSalle: Open Court Publishing Company, 1985.

Leplin, J., ed. *Scientific Realism.* Berkeley: University of California Press, 1984.

Lewontin, R.C. "Adaptation." In *Conceptual Issues in Evolutionary Biology: An Anthology,* edited by E. Sober, 234-251. Cambridge: MIT Press, 1984.

Lewontin, R.C. "The Shape of Optimality." In *The Latest on the Best: Essays on Evolution and Optimality,* edited by J. Dupre, 151-159. Cambridge: MIT Press, 1987.

Lightman, B. "Ideology, Evolution and Late-Victorian Agnostic Popularizers." In *History, Humanity and Evolution: Essays for John C. Greene,* edited by J. Moore, 285-309. Cambridge: Cambridge University Press, 1989.

Lightman, B. *The Origins of Agnosticism: Victorian Unbelief and the Limits of Knowledge.* Baltimore: Johns Hopkins University Press, 1987.

Lindberg, D.C., and R.L. Numbers. "Beyond War and Peace: A Reappraisal of the Encounter between Christianity and Science." *Church History* 55 (1986): 338-354. Reprinted in *Perspectives on Science and Christian Faith* 39 (1987): 140-149.

Lonergan, B.J.F. *Insight.* London: Longmans, Green and Company, 1957.

Løvtrup, S. *Darwinism: The Refutation of a Myth.* London: Croom Helm, 1987.

MacKay, D.M. *Behind the Eye,* edited by V. MacKay. 1986 Gifford Lectures. Oxford: Basil Blackwell, 1990.

MacKay, D.M. "Book review of *Science and Moral Priority: Merging Mind, Brain, and Human Values,* by R.W. Sperry." *Neuropsychologia* 22 (1984): 385-386.

MacKay, D.M. "Cerebral Organization and the Conscious Control of Action." In *Brain and Conscious Experience,* edited by J.C. Eccles, 422-445, 566-574. New York: Springer-Verlag, 1966.

MacKay, D.M. "Commentary: Selves and Brains." *Neuroscience* 3 (1978): 599-606.

MacKay, D.M. "Complementarity II." *Proceedings of the Aristotelian Society Supplement* 32 (1958): 105-122.

MacKay, D.M. "Complementarity in Scientific and Theological Thinking." *Zygon: Journal of Religion and Science* 9 (1974): 255-244.

MacKay, D.M. "The Health of the Evangelical Body: What Can We Learn from Recent History?" *Journal of the American Scientific Affiliation* 38 (1985): 258-265.

MacKay, D.M. *Information, Mechanism and Meaning.* Volume 1, Selected Papers. Cambridge: MIT Press, 1969.

MacKay, D.M. "The Interdependence of Mind and Brain (Letter to the Editors)." *Neuroscience* 5 (1980): 1389-1391.

MacKay, D.M. "Objectivity as a Christian Value." In *Objective Knowledge: A Christian Perspective*, edited by P. Helm, 15-27. Leicester: Intervarsity Press, 1987.

MacKay, D.M. "Our Selves and Our Brains: Duality Without Dualism." *Psychoneuroendocrinology* 7 (1982): 285-294.

MacKay, D.M. "Perception and Brain Function." In *The Neurosciences: The Second Study Program*, edited by F.O. Schmitt, G.C. Quarton, T. Melnechuk, G. Adelman and T.H. Bullock, 303-316. New York: Rockefeller University Press, 1970.

Manier, E. "Social Dimensions of the Mind-Body Problem: Turbulence in the Flow of Scientific Information." *Science and Technology Studies* 4 (1986): 16-28.

Manuel, F.E. *The Religion of Isaac Newton.* Oxford: Clarendon Press, 1974.

Marmet, P., and G. Reber. "Cosmic Matter and the Nonexpanding Universe." *Institute of Electrical and Electronics Engineers (IEEE) Transactions on Plasma Science* 17 (1989): 264-269.

Marx, J. "Homeobox Genes Go Evolutionary." *Science* 255 (1992): 399-401.

Masotti, A. "Ricci, Matteo." *Dictionary of Scientific Biography*, edited by C.C. Gillespie. New York: Charles Scribner's Sons, 1972.

Matthews, B.W. "No Code for Recognition." *Nature* 335 (1988): 244-245.

Maturana, H.R., and F.J. Varela. *Autopoiesis and Cognition: The Realization of the Living.* Dordrecht, Boston and London: D. Reidel Publishing Company, 1980.

Maynard Smith, J. "Optimization Theory in Evolution." *Annual Review of Ecology and Systematics* 9 (1978): 31-56.

Maynard Smith, J. *The Theory of Evolution.* New York: Penguin Books, 1975.

Mayr, E. "Darwin, Intellectual Revolutionary." In *Evolution from Molecules to Men,* edited by D.S. Bendall, 23-41. Cambridge: Cambridge University Press, 1983.

Mayr, E. *One Long Argument: Charles Darwin and the Genesis of Evolutionary Thought.* Cambridge: Harvard University Press, 1991.

McDonald, M.J. "Exploring 'Levels of Explanation' Concepts. Part I: Interactions between Ontic and Epistemic Levels." *Perspectives on Science and Christian Faith* 41 (1989): 194-205.

McGuire, J.E. "Boyle's Conception of Nature." *Journal of the History of Ideas* 33 (1972): 523-542.

McGuire, J.E. "Force, Active Principles, and Newton's Invisible Realm." *Ambix* 15 (1968): 154-208.

McGuire, J.E. "Newton on Place, Time, and God: An Unpublished Source." *British Journal for the History of Science* 11 (1978): 114-129.

McMullin, E. "The Conception of Science in Galileo's Work." In *New Perspectives on Galileo,* edited by R.E. Butts and J.C. Pitt, 209-257. Dordrecht: D. Reidel Publishing Company, 1978.

McMullin, E. "How Should Cosmology Relate to Theology?" In *The Sciences and Theology in the Twentieth Century,* edited by A.R. Peacocke, 17-57. Stocksfield: Oriel Press, 1981.

McMullin, E. *Newton and Matter and Activity.* Notre Dame: University of Notre Dame Press, 1978.

McTighe, T.P. "Galileo's Platonism: A Reconsideration." In *Galileo: Man of Science,* edited by E. McMullin, 365-387. New York: Basic Books, 1967.

Merton, R.K. *Science, Technology and Society in Seventeenth-century England.* Second Edition. New York: Harper and Row Publishers, 1970. Originally published in *Osiris* 4 (1938): pt. 2: 360-632.

Mesarovic, M.D. "Systems Theory and Biology—View of a Theoretician." In *Systems Theory and Biology: Proceedings, Systems Symposium 3.* Case Institute of Technology. Systems Research Center. New York: Springer-Verlag, 1968.

Mill, J.S. "Attributes." In *Essays on Ethics, Religion and Society*, edited by J.M. Robson. Toronto: University of Toronto Press, 1969.

Mill, J.S. "Theism." In *Essays on Ethics, Religion and Society*, edited by J.M. Robson. Toronto: University of Toronto Press, 1969.

Miller, J.G. *Living Systems.* New York: McGraw-Hill, 1979.

Mitchell, W.A., and T.J. Valone. "The Optimization Research Program: Studying Adaptations by Their Function." *Quarterly Review of Biology* 65, no. 1 (1990): 43-52.

Moore, J.R. "1859 and All That: Remaking the Story of Evolution-and-Religion." In *Charles Darwin, 1809-1882: Centennial Commemorative*, 167-194. Wellington: Nova Pacifica, 1982.

Moore, J.R. *The Post-Darwinian Controversies: A Study of the Protestant Struggle to Come to Terms with Darwin in Great Britain and America, 1870-1900.* Cambridge: Cambridge University Press, 1979.

More, L.T. *Isaac Newton, A Biography.* New York: Charles Scribner's Sons, 1934.

Moreland, J.P. "Theistic Science and Methodological Naturalism." in *The Creation Hypothesis: Scientific Evidence for an Intelligent Designer*, edited by J.P. Moreland, 41-66. Downers Grove: InterVarsity Press, 1994.

Mote, F.W. *Intellectual Foundations of China.* New York: Alfred A. Knopf, 1971.

Murris, H. "The Concept of the Species and Its Formation." In *Concepts in Creationism*, edited by E.H. Andrews, W. Gitt, and W.J. Ouweneel, 175-207. Welwyn Herts: Evangelical Press, 1986.

Nash, R.N. *The Concept of God.* Grand Rapids: Academe Books, 1983.

[Navarrete, D.] *The Travels and Controversies of Friar Domingo Navarrete 1618-1686*, edited by J.S. Cummins. 2 vols. Cambridge: Cambridge University Press for the Hakluyt Society, 1962.

Naylor, B.G. "Vestigial Organs are Evidence of Evolution." *Evolutionary Theory* 6 (1982): 91-96.

Needham, J. "Astronomy in Ancient and Medieval China." In *The Place of Astronomy in the Ancient World*, edited by F.R. Hodson. London: The British Academy, 1974.

Needham, J. *Chinese Astronomy and the Jesuit Mission: An Encounter of Cultures*. London: The China Society, 1958.

Needham, J. *Science and Civilisation in China*. 3 vols. Cambridge: Cambridge University Press, 1959-1962.

The New English Bible with the Apocrypha. Second Edition. New York: Oxford University Press, 1970.

Newton, I. *The Correspondence of Isaac Newton*, edited by H.W. Turnbull, J.F. Scott, A.R. Hall , and L. Tilling. 7 vols. Cambridge: Cambridge University Press, 1959-1977.

Newton, I. *Isaac Newton: Theological Manuscripts*, edited by H. McLachlan. Liverpool: University of Liverpool Press, 1950.

Newton, I. *Keynes Collection of Newton Manuscripts (MS. 2)*. Cambridge: King's College Library.

Newton, I. *Mathematical Principles of Natural Philosophy*. Translated by A. Motte and revised by F. Cajori. Second Edition. Berkeley: University of California Press, 1934.

Newton, I. *Opticks: Or, A Treatise of the Reflection, Refractions, Inflexions and Colours of Light*. Fourth Edition. New York: Dover Publications, 1952.

Newton, I. *Unpublished Scientific Papers of Isaac Newton*, edited and translated by A.R. Hall and M.B. Hall. Cambridge: Cambridge University Press, 1962.

Newton, I. *Yahuda Collection of Newton Manuscripts (MS. 14)*. Jerusalem: Jewish National and University Library.

Nicomachus. *Introductionis arithmeticae libri II*, edited by R. Hoche. Leipzig: Teubner, 1864.

North, J.D. "Astronomical Symbolism in the Mithraic Religion." *Centaurus* 33 (1990): 115-148.

North, J.D. "Proctor, Richard Anthony." In *Dictionary of Scientific Biography*, edited by C.C. Gillispie, Volume 11: 162-163. New York: Charles Scribner's Sons, 1975.

Oakley, F. "Christian Theology and the Newtonian Science: The Rise of the Concept of the Laws of Nature." *Church History* 30 (1961): 433-457.

Oakley, F. *Omnipotence, Covenant, and Order: An Excursion in the History of Ideas from Abelard to Leibniz.* Ithaca: Cornell University Press, 1984.

Obituary of M.B. Foster. *London Times* (October 16, 1959): 15.

Obituary of M.B. Foster. *Manchester Guardian* (October 16, 1959): 19.

Ogden, S. "The Meaning of Christian Hope." *Union Seminary Quarterly Review* 30 (1975): 160-163.

Oldershaw, R.L. "The New Physics—Physical or Mathematical Science?" *American Journal of Physics* 56 (1988): 1075-1081.

Orlebecke, C.J. "Donald MacKay's Philosophy of Science." *Christian Scholar's Review* 7, no. 1 (1977): 51-63.

Osawa, S., A. Muto, T. Jukes, and T. Ohama. "Evolutionary Changes in the Genetic Code." *Proceedings of the Royal Society of London* B 241 (1990): 19-28.

Osler, M.J. "Descartes and Charleton on Nature and God." *Journal of the History of Ideas* 40 (1979): 445-456.

Osler, M.J. "Eternal Truths and the Laws of Nature: The Theological Foundations of Descartes's Philosophy of Nature." *Journal of the History of Ideas* 46 (1985): 349-362.

Osler, M.J. "The Intellectual Sources of Robert Boyle's Philosophy of Nature: Gassendi's Voluntarism and Boyle's Physico-Theological Project." In *Philosophy, Science, and Religion in England, 1640-1700*, edited by R. Kroll, R. Ashcraft and P. Zagorin, 178-198. Cambridge: Cambridge University Press, 1992.

Ospovat, D. "God and Natural Selection: The Darwinian Idea of Design." *Journal of the History of Biology* 13, no. 2 (1980): 169-194.

Owen, R. *On the Nature of Limbs.* London: John Van Voorst Publishers, 1849.

Pakula, A.A., and R.T. Sauer. "Genetic Analysis of Protein Stability and Function." *Annual Review of Genetics* 23 (1989): 289-310.

Pattee, H.H., ed. *Hierarchy Theory: The Challenge of Complex Systems.* New York: George Braziller, 1973.

Patterson, C. "The Impact of Evolutionary Theories on Systematics." In *Prospects in Systematics*, edited by D.L. Hawksworth. Oxford: Clarendon Press, 1988.

Peebles, P.J.E. "Physics of the Early Universe." *Science* 235 (1987): 372.

Penrose, R. *The Emperor's New Mind: Concerning Computers, Minds, and the Laws of Physics*. Oxford: Oxford University Press 1989.

Pépin, J. "Cosmic Piety." In *Classical Mediterranean Spirituality: Egyptian, Greek, Roman*, edited by A.H. Armstrong. New York: Crossroad, 1986.

Perry, M., M. Chase, J.R. Jacob, M.C. Jacob, and T.H. Von Laue. *Western Civilization: Ideas, Politics and Society*. Third Edition. Boston: Houghton Mifflin, 1989.

Phillips, T.J. "Shroud Irradiated with Neutrons?" *Nature* 337 (1989): 594.

Philo. *Volume 4. On Abraham [De Abrahamo] and On Joseph [De Iosepho]*. Translated by F.H. Colson. 10 vols. Reprint of 1935 Edition, Loeb Classical Library. Cambridge: Harvard University Press, 1984.

Philo. *Volume 4. On Mating with the Preliminary Studies [De Congressu Quaerendae Eruditionis Gratia]*. Translated by F.H. Colson and G.H. Whitaker. 10 vols. Reprint of 1932 Edition, Loeb Classical Library. Cambridge: Harvard University Press, 1985.

Philo. *Volume 7. On the Special Laws [De specialibus legibus]*. Translated by F.H. Colson. 10 vols. Reprint of 1937 Edition, Loeb Classical Library. Cambridge: Harvard University Press, 1984.

Plantinga, A. "Methodological Naturalism." In *Facets of Faith and Science. Volume 1: Historiography and Modes of Interaction*, edited by J.M. van der Meer. Lanham: The Pascal Centre for Advanced Studies in Faith and Science/University Press of America, 1996

Plantinga, A. "On Taking Belief in God as Basic." In *Religious Experience and Religious Belief*, edited by J. Runzo and C.K. Ihara, 1-18. Lanham: University Press of America, 1986.

Plantinga, A. "When Faith and Reason Clash: Evolution and the Bible," *Christian Scholar's Review* 21 (1991): 8-32.

Plato. *The Collected Dialogues*, edited by E. Hamilton and H. Cairns. Reprint of 1961 Editions, Bollingen Series 71. Princeton: Princeton University Press, 1980.

Plato. *Opera*, edited by J. Burnet. 6 vols. Reprint of 1901-1907 Edition. Oxford: Clarendon Press, 1979-1982.

Polkinghorne, J. *Science and Creation: The Search for Understanding*. London: SPCK, 1988.

Pope John Paul II. "Address on the Work of Father Ricci in China." In *International Symposium on Chinese-Western Cultural Interchange in Commemoration of the 400th Anniversary of the Arrival of Matteo Ricci, S.J. in China*. Taiwan: Republic of China, 1983.

Popkin, R. *A History of Scepticism, From Erasmus to Spinoza*. Revised Edition. Berkeley: University of California Press, 1979.

Popper, K.R. *Conjectures and Refutations*. London: Routledge and Kegan Paul, 1963.

Popper, K.R. "Indeterminism in Classical and Quantum Physics, 1." *British Journal for the Philosophy of Science* 1 (1950): 117-133.

Popper, K.R. "Indeterminism in Classical and Quantum Physics, 2." *British Journal for the Philosophy of Science* 1 (1950): 173-195.

Popper, K.R. *Objective Knowledge: An Evolutionary Approach*. London: Oxford University Press, 1972.

Popper, K.R., and J. Eccles. *The Self and Its Brain*. Berlin: Springer International, 1977.

Proctor, R.A. "The Book of Genesis." *Knowledge* 9 (1885): 29-31.

Proctor, R.A. *Easy Star Lessons*. New York: G.P. Putnam's Sons, 1888.

Proctor, R.A. "Life in Other Worlds." *Knowledge* 11 (1888): 230-232.

Proctor, R.A. *A New Star Atlas*. London: Longmans, Green and Company, 1872.

Proctor, R.A. *The Orbs Around Us*. New York and Bombay: Longmans, Green and Company, 1902.

Proctor, R.A. *Other Worlds Than Ours*. New York: A.L. Fowle, 1870.

Proctor, R.A. "Science and Theology." *Knowledge* 6 (1884): 475-476.

"Proctor, Richard Anthony." In *Dictionary of National Biography*, edited by L. Stephen and S. Lee, 16: 419-421. London: Oxford University Press, 1959-1960.

Ptolemy. *Harmonics*. In *Greek Musical Writings II*. Translated by A. Barker. Cambridge: Cambridge University Press, 1989.

Ptolemy. *Die Harmonielehre des Klaudios Ptolemaios*, edited by I. Düring. *Göteborgs Högskolas Arsskrift* 36 (1930).

Ptolemy. *Ptolemy's Almagest*. Translated by G.J. Toomer. New York: Springer-Verlag, 1984.

Ptolemy. *Syntaxis mathematica*, edited by J.L. Heiberg. 2 vols. Leipzig: Teubner, 1898-1903.

Ptolemy. *Tetrabiblos*, edited by F. Boll and A. Boer. Leipzig: Teubner, 1957. Reprint, edited and translated by F.E. Robbins. Loeb Classical Library. Cambridge: Harvard University Press, 1980.

Putnam, H. "What is Realism?" In *Scientific Realism*, edited by J. Leplin, 140-153. Berkeley: University of California Press, 1984.

Raff, R.A., and J.C. Kaufmann. *Embryos, Genes and Evolution: The Developmental-Genetic Basis of Evolutionary Change*. New York: Macmillan Publishers, 1983.

Rao, J.K., and M. Annapurna. "Spherically Symmetric Static Inhomogeneous Cosmological Models." *Pramana* 36 (1991): 95-103.

Raup, D., and J. Valentine. "Multiple Origins of Life." *Proceedings of the National Academy of Sciences, U.S.A.* 80 (1983): 2981-2984.

Rennie, J. "Homeobox Harvest." *Scientific American* 264 (1991): 24.

Rescher, N. *Leibniz: An Introduction to His Philosophy*. Oxford: Basil Blackwell, 1979.

[Ricci, M.] *China in the Sixteenth Century: The Journals of Matthew Ricci*. Translated by L.J. Gallagher. New York: Random House, 1953.

[Ricci, M.] *Opere storiche del P. Matteo Ricci S.I.*, edited P.T. Venturi. 2 vols. Macerata: Filippo Giorgetti, 1911.

Ricci, M. *The True Meaning of the Lord of Heaven (T'ien-chu Shih-i)*, edited by E.J. Malatesta. Translated by D. Lancashire and P.H. Kuo-chen. St. Louis: Institute of Jesuit Sources, 1985.

Richards, E. "Darwin and the Descent of Woman." In *The Wilder Domain of Evolutionary Thought*, edited by D. Oldroyd and I. Langham, 57-111. Dordrecht: D. Reidel Publishing Company, 1983.

Riddiford, A., and D. Penny. "The Scientific Status of Modern Evolutionary Theory." In *Evolutionary Theory: Paths into the Future*, edited by J.W. Pollard, 1-38. New York: John Wiley, 1984.

Riddihough, G. "Homing in on the Homeobox." *Nature* 357 (1992): 643-644.

Ridley, M. *Evolution and Classification*. London: Longman, 1986.

Ridley, M. *The Problems of Evolution*. Oxford: Oxford University Press, 1985.

Rieppel, O. *Fundamentals of Comparative Biology*. Basel: Birkhauser Verlag, 1988.

Rieppel, O. "Structuralism, Functionalism, and the Four Aristotelian Causes." *Journal of the History of Biology* 23, no. 2 (1990): 291-320.

Ritter J., and K. Gründer. *Historisches Wörterbuch der Philosophie*. Band (Volume) 8. Darmstadt: Wissenschaftliche Buchgesellschaft, 1992.

Robinson, J.A.T. "The Shroud and the New Testament." In *Face to Face with the Turin Shroud*, edited by P. Jennings, 69-80. Oxford: Mowbray, 1978.

Root-Bernstein, R., and D. McEachron. "Teaching Theories: The Evolution-Creation Controversy." *American Biology Teacher* 44 (1982): 413-420.

Ross, H. *The Fingerprint of God*. Second Edition. Orange: Promise Publishing Company, 1991.

Rowbottom, A.H. *Missionary and Mandarin: The Jesuits at the Court of China*. Berkeley: University of California, 1942.

Rupke, N. "Richard Owen's Vertebrate Archetype." *Isis* 84 (1993): 231-351.

Ruse, M. *But Is It Science? The Philosophical Question in the Creation/Evolution Controversy*. Buffalo: Prometheus Books, 1988.

Ruse, M. "A Philosopher's Day in Court." In *Science and Creationism*, edited by A. Montagu, 311-342. Oxford: Oxford University Press, 1984.

Russell, B. *History of Western Philosophy*. New York: Simon and Schuster, 1945.

Russell, C.A. "Some Approaches to the History of Science." In *The 'Conflict Thesis' and Cosmology*, edited by C.A. Russell, R. Hooykaas, and D.C. Goodman. Milton Keynes: Open University Press, 1974.

Russell, E.S. *Form and Function: A Contribution to the History of Animal Morphology*. Chicago: University of Chicago Press, 1982.

Russelman, G.H.E. *Van James Watt tot Sigmund Freud—De opkomst van het stuwmodel van de zelfexpressie*. Deventer: Van Loghum Slaterus, 1983.

Ryan, M.T. "The Diffusion of Science and the Conversion of the Gentiles in the Seventeenth Century." In *In the Presence of the Past*, edited by R.T. Bienvenu and M. Feingold, 9-40. Dordrecht: Kluwer Academic Publishers, 1991.

Salthe, S.N. *Evolving Hierarchical Systems: Their Structure and Representation*. New York: Columbia University Press, 1985.

Salthe, S.N. "Self-Organization of/in Hierarchically Structured Systems." *Systems Research* 6 (1989): 199-208.

Scadding, S. "Do 'Vestigial Organs' Provide Evidence for Evolution?" *Evolutionary Theory* 5 (1981): 173-176.

Scadding, S. "Vestigial Organs Do Not Provide Scientific Evidence for Evolution." *Evolutionary Theory* 6 (1982): 171-173.

Schaffer, S. "Where Experiments End: Table-Top Trials in Victorian Astronomy." In *Scientific Practise: Theories of Stories of Doing Physics*, edited by J. Bachwald, 275-299. Chicago: University of Chicago Press, 1995.

[Schall von Bell, A.] *Lettres et Mémoires d'Adam Schall S.J. édités par le P. Henri Bernard S.J. Relation Historique*, edited and translated by P. Bornet. Tientsin: Hautes Études, 1942.

Schaller, G., H. Jinchu, P. Wenshi, and Z. Jing. *The Giant Pandas of Wolong*. Chicago: University of Chicago Press. 1986.

Schmidt-Nielson, K. "Unifying Science." *Nature* 367 (1994): 230.

Schwartz, B.I. *The World of Thought in Ancient China*. Cambridge: Belknap Press of Harvard University Press, 1985.

Scientific Creationism: A View from the National Academy of Science. Washington: National Academy Press, 1984.

Searle, J.R. "Minds, Brains and Programs." *Behavioral and Brain Sciences* 3 (1980): 417-424.

Settle, T. "The Dressage Ring and the Ballroom: Loci of Double Agency." In *Facets of Faith and Science. Volume 4: Interpreting God's Action in the World*, edited by J.M. van der Meer. Lanham: The Pascal Centre for Advanced Studies in Faith and Science/University Press of America, 1996.

Settle, T. "Fitness and Altruism: Traps for the Unwary, Biologist and Bystander Alike." *Philosophy and Biology* 8 (1993): 61-83.

Settle, T. "Van Rooijen and Mayr versus Popper: Is the Universe Causally Closed?" *British Journal for the Philosophy of Science* 40 (1989): 389-403.

Shanahan, T. "God and Nature in the Thought of Robert Boyle." *Journal of the History of Philosophy* 26 (1988): 547-569.

Simon, H.A. "The Architecture of Complexity." *Proceedings of the American Philosophical Society* 106 (1962): 467-482.

Simon, H.A. *Man and His Tools: Technology and the Human Condition.* Duijker Lecture 1981. Amsterdam: Intermediair Bibliotheek, 1981.

Simon, H.A. *The New Science of Management Decision.* Englewood Cliffs: Prentice Hall, 1960. Revised, 1977.

Simon, H.A. *The Sciences of the Artificial.* Cambridge and London: MIT Press, 1969. Second Edition. Cambridge: MIT Press, 1981.

Sloan, P. "The Question of Natural Purpose." In *Evolution and Creation*, edited by E. McMullin, 121-150. Notre Dame: University of Notre Dame Press, 1985.

Smith, R.J. "Divination in Ch'ing Dynasty China." In *Cosmology, Ontology, and Human Efficacy: Essays in Chinese Thought*, edited by R.J. Smith and D.W.Y. Kwok. Honolulu: University of Hawaii Press, 1993.

Smith, R.J., and D.W.Y. Kwok, eds. *Cosmology, Ontology, and Human Efficacy: Essays in Chinese Thought.* Honolulu: University of Hawaii Press, 1993.

Sober, E. *The Nature of Selection.* Cambridge: MIT Press, 1984.

Sober, E. *Philosophy of Biology.* Oxford: Oxford University Press and Boulder: Westview Press, 1993.

Sober, E. *Reconstructing the Past.* Cambridge: MIT Press, 1988.

Solmsen, F. *Plato's Theology*. Ithaca: Cornell University Press, 1942.

Spence, J.D. *The China Helpers: Western Advisers in China 1620-1960*. London: The Bodley Head, 1969.

Spence, J.D. *The Memory Palace of Matteo Ricci*. New York: Viking Penguin, 1984.

Sperry, R.W. "Changing Priorities." *Annual Review of Neuroscience* 4 (1981): 1-15. Revision reprinted in *Science and Moral Priority: Merging Mind, Brain, and Human Values*, by R.W. Sperry, 104-122. New York: Columbia University Press, 1983. Also reprinted as "Science, Values, and Survival." *Journal of Humanistic Psychology* 26 (1986): 8-23.

Sperry, R.W. "The Cognitive Role of Belief: Implications of the New Mentalism with Response to Howard Slaatee." *Contemporary Philosophy* 10 (1985): 2-4.

Sperry, R.W. "Commentary: Mind-Brain Interaction: Mentalism, Yes; Dualism, No." *Neuroscience* 5 (1980): 195-206. Revision reprinted in *Science and Moral Priority: Merging Mind, Brain, and Human Values*, by R.W. Sperry, 77-103. New York: Columbia University Press, 1983.

Sperry, R.W. "The Impact and Promise of the Cognitive Revolution." *American Psychologist* 48 (1993): 878-885.

Sperry, R.W. "In Defense of Mentalism and Emergent Interaction." *Journal of Mind and Behavior* 12 (1991): 221-246.

Sperry, R.W. "Mental Phenomena as Causal Determinants in Brain Function." In *Consciousness and the Brain: A Scientific and Philosophical Inquiry*, edited by G.G. Globus, G. Maxwell, and I. Savodnik, 163-177. New York: Plenum Press, 1976.

Sperry, R.W. "Mind, Brain, and Humanist Values." In *New Views on the Nature of Man*, edited by J.R. Platt, 71-92. Chicago: University of Chicago Press, 1965. Reprinted in *Science and Moral Priority: Merging Mind, Brain, and Human Values*, by R.W. Sperry, 25-44. New York: Columbia University Press, 1983.

Sperry, R.W. "A Modified Concept of Consciousness." *Psychological Review* 76 (1969): 532-536.

Sperry, R.W. "Psychology's Mentalist Paradigm and the Religion/Science Tension." *American Psychologist* 43 (1988): 607-613.

Sperry, R.W. *Science and Moral Priority: Merging Mind, Brain and Human Values.* New York: Columbia University Press, 1983.

Sperry, R.W. "Science, Values, and Survival." *Journal of Humanistic Psychology* 26 (1986): 8-23. Revised version of "Changing Priorities." reprinted in *Annual Review of Neuroscience* 4 (1981): 1-15.

Sperry, R.W. "Search for Beliefs to Live by Consistent with Science." *Zygon: Journal of Religion and Science* 26 (1991): 237-257.

Sperry, R.W. "Structure and Significance of the Consciousness Revolution." *Journal of Mind and Behavior* 8 (1987): 37-65.

Stevenson, K.E., and G.R. Habermas. *The Shroud and the Controversy.* Nashville: Thomas Nelson, 1990.

Stoeger, W. "Contemporary Cosmology and Implications for the Science-Religion Dialogue." In *Physics, Philosophy and Theology: A Common Quest for Understanding*, edited by J.R. Russell, W.R. Stoeger, and G.V. Coyne, 219-247. Vatican City: Vatican Observatory Press and Notre Dame: University of Notre Dame Press, 1988.

Strijbos, S. "The Concept of Hierarchy in Contemporary Systems Thinking—A Key to Overcoming Reductionism?" In *Facets of Faith and Science. Volume 3: The Role of Beliefs in the Natural Sciences*, edited by J.M. van der Meer. Lanham: The Pascal Centre for Advanced Studies in Faith and Science/University Press of America, 1996.

Strijbos, S. "The Concept of the Open System—Another Machine Metaphor for the Organism?" In *Facets of Faith and Science. Volume 3: The Role of Beliefs in the Natural Sciences*, edited by J.M. van der Meer. Lanham: The Pascal Centre for Advanced Studies in Faith and Science/University Press of America, 1996.

Strijbos, S. *Het technische wereldbeeld—Een wijsgerig onderzoek van het systeemdenken.* Amsterdam: Buijten and Schipperheijn, 1988.

Strijbos, S. "How can Systems Thinking Help Us in Bridging the Gap between Science and Wisdom." *Proceedings of the 37th Annual Meeting of the International Society for the Systems Sciences*, edited by R. Packham, 700-712. Sydney: International Society for the Systems Sciences, 1993.

Strong, E.W. "Newton and God." *Journal of the History of Ideas* 3 (1952): 147-167.

Stuart, C.I.J.M. "Inconsistency of the Copenhagen Interpretation." *Foundations of Physics* 21, no. 5 (1991): 591-622.

Stuart, C.I.J.M. "Negative Entropy and Entropy as Missing Information." *Physics Essays* 4, no. 2 (1991): 284-290.

Stuart, C.I.J.M. "On the Completeness of Thermodynamics." *Physics Essays* 4, no. 1 (1991): 142-143.

Stuart, C.I.J.M. and T. Settle. "Physical Laws as Knowledge and Belief." In *Facets of Faith and Science. Volume 3: The Role of Beliefs in the Natural Sciences*, edited by J.M. van der Meer. Lanham: The Pascal Centre for Advanced Studies in Faith and Science/University Press of America, 1996.

Sutter, A. *Göttliche Maschinen—Die Automaten für Lebendiges bei Descartes, Leibniz, La Mettrie und Kant.* Frankfurt am Main: Athenäum, 1988.

Swerdlow, N.M. "On the Cosmical Mysteries of Mithras." *Classical Philology* 86 (1991): 48-63.

Szentágothai, J. "Downward Causation?" *Annual Review of Neuroscience* 7 (1984): 1-11.

Talbot, A., and U. Gessner. *Systems Physiology.* New York: John Wiley and Sons, 1973.

Tamny, M. "Newton, Creation, and Perception." *Isis* 70 (1979): 48-58.

Tarán, L. *Academica: Plato, Philip of Opus, and the Pseudo-Platonic "Epinomis."* Philadelphia: American Philosophical Society, 1975.

Taub, L.C. *Ptolemy's Universe: The Natural Philosophical and Ethical Foundations of Ptolemy's Astronomy.* LaSalle: Open Court Publishing Company, 1993.

Taylor, A.E. "Plato and the Authorship of the 'Epinomis.'" *Proceedings of the British Academy* 15 (1929): 235-317.

Taylor, R.E., and R.A. Müller. "Radiocarbon Dating." In *McGraw-Hill Encyclopedia of Science and Technology.* Volume 15. Sixth Edition, 1987.

Theon of Smyrna. *Philosophi Platonici Expositio rerum mathematicarum ad legendum platonem utilium*, edited by E. Hiller. Leipzig: Teubner, 1878.

Thwaites, W.M. "Design: Can We See the Hand of Evolution in the Things It Has Wrought?" In *Evolutionists Confront Creationists: Proceedings of the 63rd Annual Meeting of the Pacific Division of the American Association for the Advancement of Science*, edited by F. Awbrey and W.M. Thwaites, Volume 1, Part 3: 212-213. San Francisco: Pacific Division of the American Association for the Advancement of Science, 1984.

Tipler, F.J. "How to Construct a Falsifiable Theory in which the Universe Came into Being Several Thousand Years Ago." In *PSA 1984: Proceedings of the 1984 Biennial Meeting of the Philosophy of Science Association.* Volume 2, edited by P. Asquith and R. Giere, 873-902. East Lansing: Philosophy of Science Association, 1984.

Tipler, F.J. "The Omega Point as Eschaton." *Zygon: Journal of Religion and Science* 24 (1989): 217-253.

Tipler, F.J. "The Omega Point Theory: A Model of an Evolving God." In *Physics, Philosophy and Theology: A Common Quest for Understanding*, edited by J.R. Russell, W.R. Stoeger, and G.V. Coyne, 313-331. Vatican City: Vatican Observatory Press and Notre Dame: University of Notre Dame Press, 1988.

Trengrove, L. "Newton's Theological Views." *Annals of Science* 22 (1966): 277-294.

Trevarthen, C.B., ed. *Brain Circuits and Functions of the Mind: Essays in Honor of Roger Sperry.* New York: Cambridge University Press, 1990.

Ulansey, D. *The Origins of the Mithraic Mysteries: Cosmology and Salvation in the Ancient World.* New York: Oxford University Press, 1989.

Van den Brom, L.J. *God alomtegenwoordig.* Kampen: J.H. Kok, 1982.

van den Wyngaert, A., ed. *Sinica Franciscana.* 3 vols. Florence: Collegium S. Bonaventure, 1936.

van der Meer, J.M. "Beliefs in Science: Taking the Measure of Methodological Materialism." In *Proceedings of the Wheaton Theology Conference 1993. Volume 2: The Relationship between Theology and Science.* Wheaton: Wheaton College Press, in press.

van der Meer, J.M. "The Struggle between Christian Theism, Metaphysical Naturalism and Relativism: How to Proceed in Science?" *Faculty Dialogue* 26 (in print).

van der Meer, J.M., ed. *Facets of Faith and Science.* 4 vols. Lanham: The Pascal Centre for Advanced Studies in Faith and Science/University Press of America, 1996.

Van Fraassen, B. "Empiricism in the Philosophy of Science." In *Images of Science*, edited by P. Churchland and C. Hooker, 245-305. Chicago: Chicago University Press, 1985.

Van Fraassen, B. *Images of Science.* Oxford: Oxford University Press, 1980.

Van Fraassen, B. "To Save the Phenomena." In *Scientific Realism*, edited by J. Leplin. Berkeley: University of California Press, 1984.

Van Till, H.J. "The Character of Contemporary Science." In *Portraits of Creation: Biblical and Scientific Perspectives on the World's Formation*, edited by H.J. Van Till, R.E. Snow, J.H. Stek, and D.A. Young, 126-165. Grand Rapids: William B. Eerdmans Publishing Company, 1990.

[Verbiest, F.] *Correspondance de Ferdinand Verbiest de la Compagnie de Jésus 1623-1688), Directeur de l'Observatoire de Pékin*, edited by H. Josson and L. Willaert. Brussels: Palais des Académies (Commission Royale d'Histoire), 1938.

von Bertalanffy, L. "Der Organismus als physikalisches System betrachtet." *Die Naturwissenschaften* 28 (1940): 521-531. Reprinted in *General System Theory*, by L. von Bertalanffy, chapter 5.

von Bertalanffy, L. *General System Theory—Foundations, Development, Applications*. Harmondsworth: Penguin Books, 1968.

von Bertalanffy, L. *Perspectives on General System Theory: Scientific-Philosophical Studies*. New York: George Braziller, 1975.

von Bertalanffy, L. *Problems of Life: An Evaluation of Modern Biological Thought*. London: Watts and Company, 1952.

von Bertalanffy, L. *Robots, Men and Minds: Psychology in the Modern World*. New York: George Braziller, 1967.

Vrba, E., and N. Eldredge. "Individuals, Hierarchies and Processes: Toward a More Complete Evolutionary Theory." *Paleobiology* 10, no. 2 (1984): 146-171.

Vrba, E.S., and S.J. Gould. "The Hierarchical Expansion of Sorting and Selection: Sorting and Selection Cannot be Equated." *Paleobiology* 12, no. 2 (1986): 217-228.

Waddington, C.H. "Fields and Gradients." In *Major Problems in Developmental Biology*, edited by M. Locke, 105-124. London: Academic Press, 1966.

Wagner, G.P. "The Biological Homology Concept." *Annual Review of Ecology and Systematics* 20 (1989): 51-69.

Weaver, R.F., and P.W. Hedrick. *Genetics*. Dubuque: Wm. C. Brown Publishers, 1989.

Webster, G. "The Relations of Natural Forms." In *Beyond Neo-Darwinism: An Introduction to the New Evolutionary Paradigm*, edited by M.W. Ho and P.T. Saunders, 193-217. London: Academic Press, 1984.

Weeks, I., and S. Jacobs. "Theological and Philosophical Presuppositions of Ancient and Modern Science: A Critical Analysis of Foster's Account." In *Creation, Nature, and Political Order in the Philosophy of Michael Foster (1903-1959): The Classic "Mind" Articles and Others, with Modern Critical Essays*, edited by C. Wybrow, 255-268. Lewiston and Lampeter, Wales: Edwin Mellen Press, 1992.

Westfall, R.S. *Force in Newton's Physics*. New York: American Elsevier, 1971.

Westfall, R.S. "Isaac Newton: Religious Rationalist or Mystic?" *Review of Religion* 22 (1958): 155-170.

Westfall, R.S. "Isaac Newton's *Theologiae gentilis origines philosophicae*." In *The Secular Mind*, edited by W.W. Wagar, 15-34. New York: Holmes and Meier, 1982.

Westfall, R.S. *Never at Rest, A Biography of Isaac Newton*. Cambridge: Cambridge University Press, 1980.

Westfall, R.S. "Newton's Theological Manuscripts," In *Contemporary Newtonian Research*, edited by Z. Bechler, 129-143. Dordrecht: D. Reidel Publishing Company, 1982.

Westfall, R.S. *Science and Religion in Seventeenth-Century England*. New Haven: Yale University Press, 1973.

White, A.D. *A History of the Warfare of Science with Theology in Christendom*. New York: n.p., 1896.

Whitehead, A.N. *Process and Reality*. New York: MacMillan Publishers, 1929. Corrected Edition, edited by D.R. Griffin and D. Sherburne. New York: Free Press, 1978. Reprint. New York: Macmillan Publishers, 1992.

Whittaker, J. "Platonic Philosophy in the Early Empire." *Aufsteig und Niedergang der Römischen Welt*. Teil II, Band 36, Volume 1 (1987): 81-123.

Whyte, L.L. "Structural Hierarchies: A Challenging Class of Physical and Biological Problems." In *Hierarchical Structures*, edited by L.L. Whyte, A.G. Wilson and D. Wilson, 3-16. New York: American Elsevier, 1969.

Whyte, L.L., A.G. Wilson, and D. Wilson, eds. *Hierarchical Structures*. New York: American Elsivier, 1969.

Willey, B. *The Seventeenth-Century Background: Studies in the Thought of the Age in Relation to Poetry and Religion.* New York: Columbia University Press, 1935.

Wilson, I. *Holy Faces, Secret Places.* New York: Doubleday, 1991.

Wojcik, J.W. "The Theological Context of Boyle's *Things Above Reason.*" In *Robert Boyle Reconsidered,* edited by M. Hunter, 139-155. New York: Cambridge University Press, 1994.

Wybrow, C. *The Bible, Baconianism, and Mastery over Nature: The Old Testament and its Modern Misreading.* New York: Peter Lang Publishers, 1991.

Wybrow, C. "Introduction: The Life and Work of Michael Beresford Foster." In *Creation, Nature, and Political Order in the Philosophy of Michael Foster (1903-1959): The Classic "Mind" Articles and Others, with Modern Critical Essays,* edited by C. Wybrow, 3-44. Lewiston and Lampeter, Wales: Edwin Mellen Press, 1992.

[Xavier, F.] *The Life and Letters of St. Francis Xavier,* edited by H.J. Coleridge. 2 vols. Fourth Edition. London: Burns and Oates, 1924.

Yandell, K. "Protestant Theology and Natural Science in the Twentieth Century." In *God and Nature: Historical Essays on the Encounter between Christianity and Science,* edited by D.C. Lindberg and R.L. Numbers, 448-471. Berkeley: University of California Press, 1986.

Yi-long, H. "Court Divination and Christianity in the K'ang-hsi Era." *Chinese Science* 10 (1991): 1-20.

Young, J.D. *East-West Synthesis: Matteo Ricci and Confucianism.* Hong Kong: Centre of Asian Studies, University of Hong Kong, 1980.

Young, R. *Darwin's Metaphor: Nature's Place in Victorian Culture.* Cambridge: Cambridge University Press, 1985.

Zanstra, H. "Is Religion Refuted by Physics or Astronomy?" *Vistas in Astronomy* 10 (1967): 1-21.

Zugibe, F.T. *The Cross and the Shroud.* New York: Paragon Press, 1988.

Zylstra, U. "Living Things as Hierarchically Organized Structures." *Synthese* 91 (1992): 111-133.

Index

Numbers in **boldface** refer to definitions

in mathematics 9
in Newton; *see:* Newton
Proctor on 37
religion as motivation 9
science serves religion 64
in the seventeenth century 135
Sperry on 209, 215
Resurrection
belief in 131
dematerialization in 127
Retrodiction barriers 60
Ricci, M. 16, 19, 21, 22, 24, 25
Ridley, M. 186
Rieppel, O. 237, 238
Ross, H. 55
Rudimentary organs 170, 176
Russell, B. 194
Russelman, G.H.E. 158

Sauer, R.T. 236
Schaffer, S. 39, 40, 42
Schall, A. 19, 21, 22, 24
Schaller, G. 183
Scherer, S. 177
Schmidt-Nielsen, K. 239
Schreck, J. (Terrentius) 21
Schrödinger, E. 103, 111
equation 103
Schwartz, B. 17
Science
comparison of MacKay and Sperry on 211
as handmaiden of religion 64
knowledge in 99, 139
limitations of 205
metaphysical implications of 97-98
and religion; *see:* Relationship between science and religion
theological presuppositions of knowledge in 146
theology of creation and knowledge in 139

unification of 251
voluntarist view of knowledge in 147
Scripture; *see:* Bible
Searle, J.R. 248
Second thermodynamic law 111
Self-assembly; *see:* self-organization
Self-organization 230
paradox of 227
Sensation
according to Plato 5
Shroud of Turin 119, 129
authenticity of 131
carbon contamination of 122
no foundation for faith 131
history of 119-21
as irreproducible event 127
as parable 130
role of presuppositions in interpretation of 129-31
production of image 128
production of secondary radiocarbon 128, 130
radiocarbon dating of 123-24
not a work of art 121
Sigonius 69
Simon, H. 244, 247, 248, 252
on complexity 250
distinction between the natural and the artificial 252
on formal hierarchy 250
on hierarchy 244, 250-51
on human dignity 249
on reductionism 248
on reductionists as pragmatic holists 248
Simplicius 6
Smith, J.D.H. 230
Smith, M. 180
Smith, R.J. 18

and scientific knowledge 139, 146
Theon of Smyrna 7, 8
Theory
predictive success of 105-7
Theory choice
criteria for 54, 55, 60
Thermodynamics
incompleteness of 111, 112
second law of 111
Thomas, R. 236
Thomson, W. (Lord Kelvin) 109
Time
as primitive concept in physics 98
reversal of 113
Tipler, F. 56
construction of falsifiable creationist models 49
Tours, Gregory of 69
Trinitarianism 67, 68, 71

Uncertainty relations
Copenhagen view of 104
of Heisenberg 104
as laws of nature 104
meaning of 104
Unification of the sciences 251
Uniformity principles 48
Unitarianism 67
Universe
center of 50
heat death of 111
higher-dimensional 57
illusion of age 49
isotropy of 51
static 52
steady state 51

van den Brom, L.
higher dimensional universe 57-59
van der Meer, J.M. 125

Van Helmont 78
Van Till, H. 54, 55
Verbiest, F. 22, 25
Verification
in cosmology 52
Vestigial organs 170, 173, 176
Victor 69
Vitalists
v mechanists 160
Voltaire 73, 89
Voluntarism/Voluntarist 137, 142, 145
on divine activity 81
and knowledge in science 147
on matter 80
theology of 138, 147
Von Bertalanffy, L. 159-62, 164, 166, 167, 245, 252
on complexity 162
on continuity 246
contrast with Dooyeweerd 251
criticism of machine models 159, 163
dynamic view of the organism 163
dynamic worldview 163-65
on formal uniformity of reality (structural uniformity) 247
Heraclitean philosophical perspective 165
on hierarchy 244
mechanists v vitalists 160
on open system 244
on reducing reality to the logical-mathematical 247, 251
on reductionism 246-47
on relation of structure and function 164
on unification of the sciences 251

Contributors

VOLUME 3:
THE ROLE OF BELIEFS IN THE NATURAL SCIENCES

JOHN BYL, Department of Mathematical Sciences, Trinity Western University, Langley, British Columbia

EDWARD B. DAVIS, Department of Science and History, Messiah College, Grantham, Pennsylvania

BERNARD LIGHTMAN, Division of Humanities, York University, North York, Ontario

MARVIN J. MCDONALD, Department of Psychology, Augustana College University, Camrose, Alberta

PAUL A. NELSON, Department of Philosophy, University of Chicago, Chicago, Illinois

TOM SETTLE, Department of Philosophy, University of Guelph, Guelph, Ontario

SYTSE STRIJBOS, Department of Philosophy, Free University, Amsterdam, The Netherlands

C. (IAIN) J. M. STUART (deceased), Department of Applied Sciences in Medicine, University of Alberta, Edmonton, Alberta

LIBA TAUB, Associate Curator for the History of Astronomy Collection, The Adler Planetarium, Chicago, Illinois

THADDEUS TRENN, Institute for the History and Philosophy of Science and Technology, Victoria College, University of Toronto, Toronto, Ontario

JITSE M. VAN DER MEER, Department of Biology and Director of the Pascal Centre, Redeemer College, Ancaster, Ontario

RICHARD S. WESTFALL, Professor Emeritus, Department of Philosophy of Science, University of Indiana, Bloomington, Indiana

DAVID L. WILCOX, Department of Biology, Eastern College, St. Davids, Pennsylvania